Arthur Lachman

The spirit of organic chemistry

An introduction to the current literature of the subject

Arthur Lachman

The spirit of organic chemistry
An introduction to the current literature of the subject

ISBN/EAN: 9783337203931

Printed in Europe, USA, Canada, Australia, Japan

Cover: Foto ©berggeist007 / pixelio.de

More available books at **www.hansebooks.com**

THE SPIRIT

OF

ORGANIC CHEMISTRY

AN INTRODUCTION TO THE CURRENT
LITERATURE OF THE SUBJECT

BY

ARTHUR LACHMAN

(B.S., CALIF.; PH.D., MUNICH)

PROFESSOR OF CHEMISTRY IN THE UNIVERSITY OF OREGON

WITH AN INTRODUCTION BY

PAUL C. FREER, M.D., PH.D.

PROFESSOR OF GENERAL CHEMISTRY IN THE
UNIVERSITY OF MICHIGAN

New York
THE MACMILLAN COMPANY

LONDON: MACMILLAN & CO., LTD.

1899

ERRATA

Page 50, line 4, read " of distance " in place of " on distance."

Page 129, line 4 below Chart I., read " formula " in place of " formulæ."

Page 131, corrected cut : —

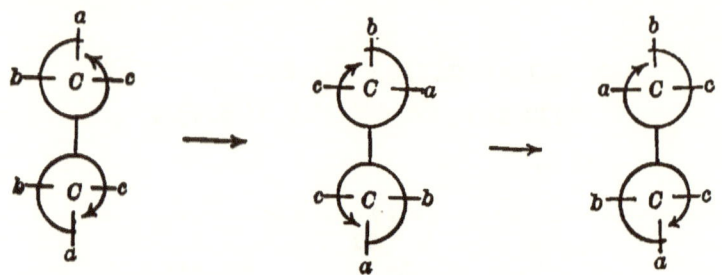

Page 146, footnote 2, omit c in *Monatsh.*

Page 162, corrected cut : —

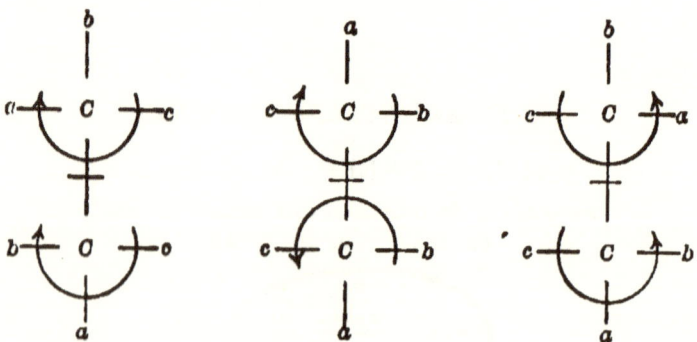

Page 173, line 12, 2d formula, read " $C_6H_5 - C = NOH$ " instead of " $C_6H_5 = C = NOH$."

Page 182, line 10 from bottom, read " NX " instead of " NX_2."

Page 202, line 2 from bottom, 2d part of formula, read " $(CH_3)_3N - O$ " instead of " $(CH_3)_3 - N \longrightarrow O$."

Page 209, line 13, read " $KHSO_4$ " instead of " K_2HSO_4."

PREFACE

THIS book is intended primarily as a supplement to text-books of organic chemistry. Text-books tell us the important facts of a science and the conclusions to which these have led; their aim is to present the subject and its principles as a complete whole — their emphasis is laid upon unity and coördination. But a science is a living, growing thing, which at any given moment is far from being unified and coördinated. The beginner, coming upon the ten thousand pages which mark the annual growth of organic chemistry, cannot but be bewildered. Theories stated conclusively by his text-book he finds assailed and refuted; facts unmask themselves which he is tempted to regard as impossible. Why this ceaseless activity? Are the text-books wrong? Are these ten thousand pages of vital importance, and must one read them all if he hopes to keep up with the march of science? Why is there so much "research"? To all of these questions the student finds no answer ready to hand; this little compilation aims to satisfy his curiosity.

The scientific investigator must test his theories by the light of special facts. His general ideas resolve themselves into a number of component ideas; research, in the last analysis, consists of a series of *detailed problems*, more or less extended.

A general principle of organic chemistry, therefore, must be examined by the side of *definite* experiments — an experiment is the very essence of definiteness. The present state of its development is best discerned by following its historic evolution, by a study of its origin and of its career. A number of the chief problems of the science have been taken up in this manner, and will be found to answer many of the questions that perplex the rising chemist. Problems which have had no history, which have been established in a single masterly research, manifestly do not fall within the scope of this collection; they are sufficiently treated in the text-books. No great demand has been made upon previous familiarity with the subject. A certain danger lies in the attempt to bring the topics up to date; but wherever possible this risk has been incurred. To avoid confusion, formulæ antedating present structure-theories have been transposed into their modern equivalents.

A word of apology may be necessary for the authorship of this somewhat critical volume. There can be no doubt that it would grace the pen of a master better than that of a tyro. But the leaders in the van of progress do not heed the wants of the struggling columns in their rear — and so we must needs help ourselves. The habit of browsing among the earlier numbers of our chemical journals, originally formed for self-improvement, has been the immediate suggestion for the present work. The detailed study of treatises of bygone periods has been its own reward; and this little volume will fail of its purpose if its readers are not induced to seek out the sources of our present knowledge.

The materials employed during the preparation of the various chapters are sufficiently indicated in the body of the work. Wherever use has been made of existing summaries the fact has been mentioned; but the total treatment of each subject has always been evolved independently. The attempt has been made to quote the original authorities for all important statements; and unsupported comments and criticisms must be credited, or, it may be, blamed, to the author. The author must also bear the responsibility for his choice of material; the main object of the book being didactic, the individual chapters are by no means to be regarded as bibliographies.

Aside from the main object just stated, the method of treatment here adopted may serve other desirable ends. In the first place, the personal element in science would appear to be far too much neglected. Nothing brings home to the student the living, growing nature of his subject more than to be familiar with the names of those who have brought it to its present stage, or perchance to hear his instructors tell of their ways in the laboratory and the lecture-room. The share of credit due to each investigator in the various topics under consideration has been strongly emphasized.

Secondly, students are not sufficiently encouraged to read the current literature; and its vastness proves an effective damper if no guide to its mazes be at hand. As most of the subjects selected here are still open questions, as the names of the investigators engaged upon them are prominently put forward, and as the journals in which these chemists are accustomed to publish their results are invariably quoted, no difficulty should be experienced in following up *these* topics, at any rate. Once the habit of perusing the journals is formed, the ability to discern the gold among the dross comes rapidly. The amount of actually important work published is but a small fraction of the total. The conception of research given above 'will afford an unfailing touchstone in doubtful cases.

In the third place, students seem to select their universities for every possible reason other than that certain men are engaged upon certain investigations there. There can be no doubt that this condition is undesirable. It seems to be due to a lack of appreciation of the importance and meaning of research work, as well as to ignorance concerning the particular activities of the instructors. To a certain extent the remedy is offered herewith.

Finally, the essays which make up this volume may be regarded as a small contribution to the history of Science. It is to be regretted that organic chemistry is commonly regarded as a labyrinthine specialty, not only among brother scientists, but among brother chemists as well. Organic chemistry is the physiology of molecular science. No branch of human activity can vie with its rounded symmetry and completeness, can reduce so large a percentage of its facts to law and order, or can approach nearer to the ultimate mystery of matter. And so the study of systematic progress and development which it offers should become an open book to every man of science.

The compilation of this work was begun in January, 1897, at Ann Arbor, and completed in October, 1898, in Eugene. The author must not fail to express his gratitude to Drs. Freer and Sherman of the University of Michigan and to Dr. Solomons of the University of Wisconsin for their patient and vigorous criticism during the preparation of the greater part of the manuscript.

ARTHUR LACHMAN.

EUGENE, OREGON,
    October, 1898.

CONTENTS

CHAPTER VIII

CHAPTER IX

INTRODUCTION

THE GROWTH OF THE SCIENCE OF ORGANIC CHEMISTRY

THE magnificent science of modern Organic Chemistry, with its countless compounds, — formed of but few elements, yet varying in kaleidoscopic array, at one time showing us volatile oils, at another infusible solids, embracing brilliant-hued dyes, powerful poisons, beneficent medicines, sweet sugars, and bitter alkaloids, contrasting the most agreeable with the most nauseating odors; with its varying reactions, dealing at one time with acids, at another with bases, and yet again with neutral substances; with the extreme mutability of the bodies which come within its realm, — fascinates with its possibilities, while it appalls with its vastness. The student stands discouraged at the very threshold. How can he ever hope to master the general classification, let alone the minor details, which must become a part of his very being, if he too wishes to do his share, however small, toward completing and rounding out the still unfinished structure? The answer is plain : he can do this only by comprehending the *spirit of the science*, by learning its great theories, not as mere mnemonic efforts, but as the result of a development for which many of the most earnest and acute minds known to the history of science have fought and toiled. The whole subject has in the past been in a

state of flux, it is so at present, and its point of con-
gelation is still in the far distant future. None of
the great generalizations of to-day have come to us
through the efforts of one man; they are, on the contrary, the result of a gradual growth, in each step of
which the mental acumen of some investigator, perhaps long since dead, can be seen, and each great
research of to-day is built upon some perhaps equally
great one of yesterday. We can no more comprehend the present structural chemistry without
recalling Berzelius, Wöhler, Liebig, Dumas, Gerhardt, Kolbe, or Laurent, than we can the Roman
Empire without recognizing the part played in its
foundation by Cæsar and Augustus.

When in 1828 Wöhler demonstrated the formation of urea from ammonium cyanate, he at one stroke
broke down the barrier which at that time existed
between mineral and organic chemistry. Previously
it had been believed that all substances belonging in
the domain of the latter were of necessity the product of a peculiar " vital force," which could be made
subservient to man only in so far as he might himself
be able to construct a living organism; and this in
itself, with the given hypothesis, was a manifest
impossibility. Now all was changed; a product of
vital metabolism, and, indeed, a most important one,
had been produced in the laboratory, and from so-
called inorganic constituents. Henceforth the two
branches of the science were to travel side by side,
the great generalizations of the one aiding in the
development of the other, until now it can truly be
said, *no one can understand the spirit of inorganic
without being thoroughly conversant with that of*

organic chemistry. The attention of chemists was turned from purely analytical lines, from questions of identification, correlation, and classification of mineral substances to the great field of synthesis, involving the creation of multitudes of new and interesting substances. The expression "nothing is new in this world" cannot surely be applied to organic chemistry, for in this chosen field of so many laborers we see daily, and even hourly, new individuals brought forth and christened in the laboratory, and the like of these the eye of man never beheld before.

Although this new field of science can point to an unbroken record of progress from its very beginning, although it has enriched the realm of human knowledge with many and important discoveries, it has nevertheless been the scene of most persistent strife, often carried on with an acrimony and bitterness that shocks our modern sense of professional courtesy. Its triumphs have called for sacrifices of human lives and broken hearts, but the obstinacy of past conflicts has taught us to respect the views of others, and to seek for a verification of our own, not in dogmatic discussion, but in careful and logical experimentation, painstaking in the minutest detail, each part a building stone from which the final edifice is to be constructed. Following each other in rapid succession, there arose theory after theory, each having for its aim the explanation of the *structure* of the organic bodies so far as known; each strong in that it was founded upon some experimental facts ; each weak in that it could not embrace in one view the *whole* of the field it sought to cover;

each in turn to be overthrown, but each leaving its germ of truth to be utilized by its successors. First, greatest, and most obstinately maintained, the dualistic theory of Berzelius, finally giving way to the substitution doctrine of Gerhardt and Laurent; then the radicle theory of Liebig and Wöhler, subsequently to be transformed and absorbed by the theory of types—all views finally brought into contact at once, so that Laurent, in an attempt to clear up the resulting chaos, sarcastically, and yet rather sadly, calls attention to the fact that for such a simple substance as acetic acid, as many as eleven different structural formulæ were at one time proposed.[1] Indeed, this confusion was inevitable so long as at the same time an equal confusion reigned in a kindred topic of the science—the determination of atomic weights. Not until Cannizzaro, by his masterly revival of Avogadro's hypothesis, taught chemists to use a consistent and logical method of determining atomic weights; not, then, until investigators were able to know *how many atoms are included in each molecule*, were they able to determine properly the relative positions which those atoms bear to each other in any given chemical compound. From the remnants of the past conflicts much has been taken. The dualistic theory of Berzelius we still find, shorn of its dogma and its inflexibility, in a newer and more reasonable view of the influence of positive and negative constituents on the character of chemical compounds; Liebig's and Wöhler's conception of compound radicles, however

---

[1] Laurent, *Chemical Method*, p. 23.

greatly modified, is still a part of modern chemistry; the substitutions of Dumas are as true to-day as they were fifty years ago—all, however, under the unifying influence of Cannizzaro's revival, are brought into one great stream by Williamson's and Kekulé's theory of valence and of structural chemistry. It is this advance that has made the great strides of recent years possible; it is this that has produced artificial indigo and caffeïn, cleared up the dark mystery of the sugar group, given us a concise and clear understanding of the derivatives of benzene and of the terpenes, made possible the development of the aniline and azo-dye industries, enriched medicine with antipyrine, phenacetine, antifebrine, and in countless ways has contributed both to the health and the happiness of mankind. To-day we can scarcely imagine ourselves without these adjuncts of modern civilization, but if we do picture to ourselves what the world would be were we deprived of them, we must give all credit to the master minds that made these things possible. Truly, in no other field of human progress can it be said with equal truth that the investigations of scientific men, working for the cause of science alone and without thought of pecuniary gain, must inevitably have preceded the commercial application of their discoveries, for out of the chaos of the past have sprung the industries of the present.

No student of organic chemistry can afford to neglect the historic side of the question; no one can claim to have absorbed its spirit unless he understands how the knowledge of the present has grown from the toil of the past, and none may expect to create for himself unless he understands the methods and views of his predecessors. In so doing he must give his whole soul to the task before him. Science is a stern mistress, who gives of the best within her only to those who follow her unflinchingly, however difficult the task, however remote the prospect of pecuniary gain or of self-aggrandizement, their sole hope being that they too may add to mankind's knowledge of truth, so that future generations may profit by the sacrifices of the present. This has been the spirit of the past; it must also be the spirit of the present and of the future. Science is moving onward, swiftly, relentlessly, unflinchingly — no half-hearted followers for her — the weak fall by the wayside, there is no place for those who have not the patience to acquire the necessary knowledge, the strong press forward in fierce rivalry, each striving for the ultimate goal — a perfect human knowledge, by which from any given premises the logical conclusion may be drawn with unerring accuracy. *Finis coronat opus.*

PAUL C. FREER.

ANN ARBOR, 1899.

THE SPIRIT OF ORGANIC CHEMISTRY

•

CHAPTER I

THE CONSTITUTION OF ROSANILINE

NOT very long ago, the problem of determining the constitution of rosaniline, pararosaniline, and their derivatives was by far the most important one in the whole domain of technical organic chemistry. In 1857 W. H. Perkin began the industrial preparation of mauve, the first so-called "aniline-dye." The phenomenal success attendant upon his efforts roused a perfect whirlwind of competitive enthusiasm. Everybody who could devise a new oxidizing agent immediately tried it on aniline, obtained a new dyestuff, and forthwith patented his process. Chief among these aniline dyes was *fuchsine*, the storm-centre itself. In France a tremendous monopoly arose, which attempted to force its way into England and Germany, but came to grief thereby. The struggles of its competitors called forth feverish activity on the part of the skilled chemists of these countries; for in the secret of the exact composition of fuchsine lay the method for its most successful preparation. This secret was completely bared in 1878, and the expected results were not long in arriving. At almost the same moment, however, the

1

commercial importance of the fuchsines began to wane, for the azo-dyes commenced their triumphant career, and to-day have largely replaced their earlier competitors.

It is to A. W. Hofmann that we owe most of our knowledge of the methods for preparing fuchsine derivatives. The first formula assigned to the mother-substance was $C_{24}H_{10}N_2O_2$. Hofmann showed this to be wrong; the fuchsine base, *rosaniline*, is represented by the composition $C_{20}H_{21}N_3O$.[1] The aniline first employed in the manufacture of the dyes boiled between 180°–220°; manifestly it was very impure. With the progress of technical skill came the ability to prepare *pure* aniline; but it was found that the oxidation of pure aniline gave no fuchsine at all. Hofmann was called upon to investigate this curious result. He at once concluded that the formation of fuchsine was due to the *homologues* of aniline, which certainly must be present in the crude product. Their isolation from this mixture seemed hopeless; but fortunately Hofmann possessed a small quantity of toluidine, the remnant of an earlier investigation on toluene. The oxidation of toluidine, however, gave no fuchsine. It was not until he oxidized a *mixture* of aniline and toluidine that Hofmann[2] was rewarded by a most brilliant red — the para-rosaniline of to-day. Technically the problem was solved; scientifically it received a great impetus, for the mysterious twenty carbon atoms in rosaniline were seen to consist of six from the aniline and twice seven from two molecules of toluidine.

[1] *Proc. Roy. Soc.* **12**, 2 (1862).   [2] *Ibid.* **13**, 490 (1864).

Hofmann's further investigations showed that rosaniline contained replaceable hydrogen atoms; for by action of methyl or ethyl iodides, or of aniline, the radicals methyl, ethyl, and phenyl could be introduced : phenyl three times,[1] the others six times. This led him to the conclusion that the number of replaceable hydrogen atoms was *three;* the hexamethyl and ethyl compounds he regarded as quaternary ammonium bases, phenyl not forming quaternary compounds. Other items of importance for the determination of the constitution of the mother-substance were also discovered by Hofmann; viz. that the *salts* of rosaniline, the actual dyes, contained *no* oxygen, but were derived from the compound $C_{20}H_{19}N_3(C_{20}H_{21}N_3O - H_2O)$.   Reduction of the base led to the corresponding *leuco*-compound,[2] $C_{20}H_{21}N_3$. These various facts led him to consider rosaniline as a triamine with three free hydrogen atoms.   Kekulé[3] expressed this conception by the following formula :

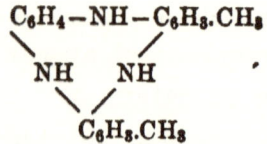

This formula did not obtain general recognition; from occasional footnotes it is evident that Hofmann later looked upon rosaniline as an azo-compound;[4] and this latter view was shared by many others.   In

---

[1] *Proc. Roy. Soc.* **13**, 6 (1863).

[2] Dyestuffs are very frequently easily reducible to *colorless* substances (λευκός, white). These leuco-compounds, in turn, are readily oxidized to the dyestuff ; indigo white is a typical example.

[3] *Lehrbuch*, III, 672.   [4] *Ber. d. chem. Gesell.* **5**, 472 (1872).

addition, certain facts did not agree with Kekulé's formula. Hofmann[1] had found that nitrous acid acted upon rosaniline with formation of a new base. Caro and Wanklyn[2] showed this base to be a *diazo-compound*; on boiling with water it yielded a compound $C_{20}H_{16}O_3$, apparently rosolic acid. It now became of importance to settle the constitution of rosolic acid. Caro[3] found that it could be prepared by a process exactly analogous to that employed with rosaniline; viz. by oxidation of a mixture of phenol and cresol, *i.e.* of a benzene and a toluene derivative. On the other hand, it was by this time well established by various investigators that rosolic acid or rosaniline could be obtained from benzene derivatives alone, provided oxalic acid, carbon tetrachloride, or other simple carbon compounds be present. Evidently, the extra carbon atoms of the toluene derivatives, or of the simple compounds mentioned, are necessary for *joining* the benzene nuclei. Taking this into account, as well as the view expressed by Hofmann that three *typical* hydrogen atoms are present in the molecule, Caro and Wanklyn ascribed the following formulæ to rosaniline and rosolic acid:

$$C_2 \begin{cases} NH.C_6H_5 \\ NH.C_6H_5 \\ NH.C_6H_5 \\ H \end{cases} \qquad C_2 \begin{cases} O.C_6H_5 \\ O.C_6H_5 \\ O.C_6H_5 \\ H \end{cases}$$

At about this time (1869), Rosenstiehl[4] found that different rosanilines were obtained according to

[1] *Proc. Roy. Soc.* 12, 13 (1862).   [2] *Ibid.* 15, 210 (1866).
[3] *Ztschr. Chem.* 1866, 563.
[4] *Jahresbericht*, 1869, 693; cf. *Ber. d. chem. Gesell.* 9, 441 (1876).

whether ortho- or para-toluidine were employed for the purpose. These new facts did not collide with the then current theories, however, and Rosenstiehl made no effort to capitalize them. This was reserved for the future.

Caro continued his researches on the relation between rosolic acid and rosaniline, and in 1875[1] showed (in conjunction with Graebe) that both of these substances must be regarded as derivatives of a hydrocarbon containing twenty carbon atoms. In many respects rosolic acid behaved as a quinone: it added various bisulfites, ammonia, even hydrochloric acid; it could be reduced to a leuco-compound; with acetic anhydride it gave a triacetate. These facts Caro and Graebe attempted to condense into the formula:

$$(HO)C_6H_8 < {CH_2 \cdot C_6H_4{=}O \atop CH_2 \cdot C_6H_4{=}O}$$

Rosaniline was given the analogous constitution:

$$(NH_2)C_6H_8 < {CH_2 \cdot C_6H_4{=}NH \atop CH_2 \cdot C_6H_4{=}NH}$$

These formulæ, however, failed to account for the chief series of reactions connecting the two compounds: the formation of a diazo-compound from rosaniline, which then furnished rosolic acid. Caro and Graebe fully realized this discrepancy, but straddled it with the lame assumption that the imido-groups of rosaniline could behave like amido- groups towards nitrous acid.

Similar relationships between the rosolic acid and rosaniline compounds had been discovered by Dale

---

[1] *Ann. Chem.* (Liebig), **179**, 184 (1875).

and Schorlemmer.[1] But just at this time Emil and Otto Fischer[2] had accidentally been led to an investigation of diazo-rosaniline; and in a brief space[3] they had the matter pretty definitely cleared up.

By applying the ordinary reactions for the complete elimination of the diazo-complex from aromatic compounds, and its replacement by hydrogen, they obtained the fundamental hydrocarbon, $C_{20}H_{18}$, of which rosaniline is the derivative. Considerable difficulty was experienced in determining the identity of this hydrocarbon. It was not until the hydrocarbons resulting from the supposedly isomeric rosanilines of Rosenstiehl were investigated, that the question was settled. Rosenstiehl had ascribed to the base prepared from aniline and ortho-toluidine the name *rosaniline*, to that from para-toluidine *pseudo-rosaniline*. On reducing this pseudo-rosaniline to the leuco-base, diazotizing, and boiling with alcohol, the Fischers obtained a hydrocarbon, $C_{19}H_{16}$, which proved to be the well-known triphenylmethane:

$$HC \cdot (C_6H_5)_3$$

The hydrocarbon obtained from *technical* rosaniline, $C_{20}H_{18}$, was at once regarded as tolyl-diphenylmethane, and the supposition clinched by experimental verification.

Having thus by analysis ascertained that rosaniline (therefore also rosolic acid) is a derivative of triphenylmethane and its homologues, the Fischers proceeded to synthetically prepare rosaniline com-

---

[1] *J. Chem. Soc.* 1873, 434; 1879, 148, 562.

[2] *Ber. d. chem. Gesell.* 9, 891 (1876).

[3] *Ann. Chem.* (Liebig), 194, 242 (1878).

pounds from these hydrocarbons.   First a word concerning the revision of nomenclature necessitated by these researches.   Technical rosaniline and rosolic acid were found to be derived from tolyl-diphenylmethane; in view of the practical convenience in preserving these designations, it was decided to distinguish the lower homologue of rosaniline as *para*-rosaniline, to recall its formation from aniline and paratoluidine.   The corresponding rosolic acid (the triphenylmethane derivative) had been known for some time as " Aurine"; this name was retained.

Triphenylmethane is readily converted into trinitrotriphenylmethane.   Two methods lead from this substance to pararosaniline.   Direct reduction gives triamido-triphenylmethane, or leuco-pararosaniline; like all leuco-compounds, this is converted by careful oxidation into the dyestuff pararosaniline itself. The second method proceeds as follows : trinitro-triphenylmethane is oxidized to the corresponding carbinol by chromic acid :

$$HO . C(C_6H_4NO_2)_3$$

Trinitro-triphenylcarbinol is reduced by zinc dust and acetic acid to the amido-compound, pararosaniline.

The results obtained by the Fischers may be summarized as follows :

(1) Leuco-pararosaniline is the triamido-derivative of triphenylmethane, in which the amido-groups are evenly distributed among the three benzene nuclei. Its formation from trinitro-triphenylmethane, on the one hand, and from paratoluidine and aniline, on the other, show this conclusively.

(2) Pararosaniline and its leuco-compound are to each other as triphenylcarbinol is to triphenylmethane; this is shown by the direct formation of the dyestuff from the trinitro-carbinol.

This would lead to the following formula :

$$NH_2 . C_6H_4 \diagdown \quad \diagup C_6H_4 . NH_2$$
$$NH_2 . C_6H_4 \diagup C \diagdown OH$$

for *free* pararosaniline.   The salts of this base, however, do not contain oxygen, being derived from $C_{19}H_{17}N_3$ (pararosaniline less water).   The " color-base " would, therefore, probably have the following constitution :

$$NH_2 . C_6H_4 \diagdown \quad \diagup C_6H_4$$
$$NH_2 . C_6H_4 \diagup C \diagdown NH$$

This formula explains the easy addition of water, as well as the reduction to leuco-pararosaniline.   It also explains a much older reaction of the rosanilines. Müller[1] had shown that the rosaniline salts add hydrocyanic acid to form very stable compounds, containing *three* amido-groups.   The Fischers ascribe to hydrocyanpararosaniline the constitution :

$$NH_2 . C_6H_4 \diagdown \quad \diagup C_6H_4 . NH_2$$
$$NH_2 . C_6H_4 \diagup C \diagdown CN$$

The position of the amido-groups in the benzene rings still remains to be discussed.   An observation of Caro and Graebe's[2] determined the situation of two of them.   Aurine,[3] when superheated with water,

---

[1] *Ztschr. Chem.* **1866**, 2.

[2] *Ber. d. chem. Gesell.* **11**, 1348 (1878).

[3] Cf. above.

gives p-dioxy-benzophenone, therefore contains two of its hydroxyls in para position to the methane carbon atom :

$$CO \begin{cases} {}^{\diagup}C_6H_4OH \text{ (p)} \\ {}_{\diagdown}C_6H_4OH \text{ (p)} \end{cases}$$

As the hydroxyls of aurine correspond to the amido-groups of pararosaniline, two of the latter must be in para position. The third amido-group is also in para position.[1] The condensation of benzaldehyde with aniline to form *di*amido-triphenylmethane :

$$C_6H_5 . CHO + 2 C_6H_5 . NH_2 = C_6H_5 . CH < {}^{C_6H_4 . NH_2}_{C_6H_4 . NH_2} + H_2O$$

occurs in para position, since by first converting into the dioxy-triphenylmethane this substance also yields p-dioxy-benzophenone. Now if instead of benzaldehyde p-nitro-benzaldehyde be employed in the above reaction, and the resulting compound reduced, pararosaniline is the final product ; therefore the third amido-group occupies the para position.

These researches of E. and O. Fischer contained but two experimental gaps, which have since been filled in by E. Fischer and Jennings.[2] These latter have proved that pararosaniline is directly converted into triphenylcarbinol by the diazo-reaction, thus giving the reverse of that process whereby triphenyl-carbinol is converted into pararosaniline by nitration and subsequent reduction. At the same time it was also shown that hydrocyanpararosaniline has the constitution ascribed to it above, since (by means of

[1] E. and O. Fischer, *Ber. d. chem. Gesell.* 13, 2204 (1880).

[2] *Ber. d. chem. Gesell.* 26, 2225 (1893).

the diazo-reaction) it is converted into triphenyl acetonitrile

$$(C_6H_5)_3C-CN$$

and therefore contains the cyanogen group attached to carbon.

Of late, owing to the curious circumstance that the free rosanilines are *colorless*, whereas their salts, the fuchsines, are intensely colored, speculation has been rife as to the cause thereof. These have been mostly concerned with the constitution of the *salts*, though very recently the bases themselves have been drawn into the discussion. Nietzki [1] (1889) assigned a quinone-like structure to the fuchsines :

$$\begin{matrix} C_6H_4NH_2 \\ C_6H_4NH_2 \end{matrix} > C = C_6H_4 = NH \cdot HCl$$

this differed but little from the one proposed by E. and O. Fischer :

$$\begin{matrix} C_6H_4NH_2 \\ C_6H_4NH_2 \end{matrix} > C - C_6H_4 - NH \cdot HCl$$

viz. by just the difference between the old and the new quinone formulæ. In view of the preference which has arisen during the past few years for quinone formulæ for colored substances, this formula of Nietzki is commonly accepted. Rosenstiehl [2] (1893), indeed, advanced another possibility :

$$(C_6H_4NH_2)_3C \cdot Cl$$

regarding fuchsine as a *hydrochloric acid ester*, and not as the hydrochlorate of a base. But Miolati [3]

---

[1] *Chemie der organischen Farbstoffe*, p. 117 (3d ed.).
[2] *Bull. Soc.* [3], 9, 117 (1893).
[3] *Ber. d. chem. Gesell.* 26, 1788 (1893).

has found that the fuchsines are electrolytes, and therefore salts ; this excludes[1] Rosenstiehl's formula.[2]

The formula of pararosaniline itself has been somewhat modified by Weil[3] (1895). Pararosaniline is a strong base, while triamido-triphenylmethane is a very weak one. If both of these compounds have analogous structures, it is very strange that the hydroxyl derivative should be much the more basic of the two. Weil therefore suggests that pararosaniline must have a different structure than its mother-substance, its basic properties indicating the presence of a pentavalent nitrogen atom. Of the various possibilities which incorporate this idea, Weil's experiments show the following to be by far the most probable :

$$(C_6H_4NH_2)_2C \cdot C_6H_4 \cdot NH_3$$
$$\diagdown_O \diagup$$

Further details have been promised, and will be awaited with interest.

The history of rosaniline has been comparatively free from complications, one single research having given us an almost complete insight into its constitution. The pivotal point, of course, was the recognition of triphenylmethane as the backbone of the group. The method here exemplified, of building down to a known substance, ascertaining the original structure therefrom by induction, and verifying this induction by synthesis from that familiar compound, typifies the mental attitude of the scientist toward his problems.

---

[1] Cf. also Fischer and Jennings, *l.c.*

[2] Cf. Rosenstiehl, *Bull. Soc.* [3], 17, 373 (1897), for contradictory evidence.   [3] *Ber. d. chem. Gesell.* 28, 205 (1895).

CHAPTER II

PERKIN'S REACTION

THE condensation of aromatic aldehydes with salts of aliphatic acids, commonly known as Perkin's reaction, has had a very curious and instructive history. The reaction is extremely important to the synthetic chemist, and has contributed in no small degree to our knowledge of chemical processes.

In 1868, W. H. Perkin[1] described a synthesis of coumarine, consisting in the action of acetic anhydride on the sodium salt of salicylic aldehyde. This choice of materials was determined by the fact that coumarine is decomposed into acetic and salicylic acids on fusion with caustic alkalies. Perkin explained this reaction as follows: The first step consists in the formation of acetyl-salicylic aldehyde:

$$C_6H_4 < {}^{CHO}_{ONa} + {}^{CH_3CO}_{CH_3CO} < 0 = C_6H_4 < {}^{CHO}_{0-COCH_3} + CH_3COONa$$

The aldehyde then loses water and forms coumarine:

$$C_6H_4 < {}^{CHO}_{0-COCH_3} = C_6H_3 < {}^{CO}_{COCH_3} + H_2O$$

[1] *J. Chem. Soc.* **1868**, 53; *Ann. Chem.* (Liebig), **147**, 229 (1868).

The reasons which led Perkin to this conclusion were:

First. Acetyl-salicylic aldehyde is actually formed when the reagents employed above are brought together in cold ether solution.

Second. As the dehydration is caused either by the excess of acetic anhydride, or by the temperature attained (which did not exceed 300°), it is not likely that the acetyl group is involved, especially as acetic acid is a product of decomposition (on fusing with alkalies).

Third. Coumarine does not contain an acetyl group attached to *oxygen*, since it does not yield acetic acid on treatment with aqueous alkalies.

Fourth. Coumarine contains no aldehyde group, since it possesses no aldehyde characteristics.

However, this view was abandoned shortly. In the first place, Perkin[1] examined the behavior of acetyl-salicylic aldehyde, and found that it could not be made to form coumarine unless sodium acetate be present. He was unable to explain why. And secondly, the coumarine formula given by Perkin was sharply criticised by Fittig,[2] who pointed out that it was hardly likely for the benzene ring to give up hydrogen in such a manner, and that independently of this the formula did not agree with the properties of coumarine and coumaric acid. Fittig recalled a reaction discovered by Bertagnini[3] and explained by Baeyer,[4] in which cinnamic acid is

[1] *J. Chem. Soc.* 1868, 181; *Ann. Chem.* (Liebig), 148, 208 (1868).
[2] *Ztschr. Chem.* 1868, 595; *Ann. Chem.* (Liebig), 153, 358 (1870).
[3] *Ann. Chem.* (Liebig), 100, 125 (1856).
[4] *Ann. Chem.* (Liebig), Supp. 5, 81 (1867).

formed from benzaldehyde and acetyl chloride. Baeyer showed that the first step in this reaction is à condensation :

$$C_6H_5CHO + CH_3COCl = C_6H_5CH=CHCOCl + H_2O$$

The cinnamyl chloride subsequently reacts with water :

$$C_6H_5CH=CHCOCl + H_2O = C_6H_5CH=CHCOOH + HCl$$

and cinnamic acid is the result. Fittig offered an analogous explanation of Perkin's reaction. The aldehyde group condenses with half of the acetic anhydride, the other half replaces the sodium atom by hydrogen :

$$C_6H_4 < ^{ONa}_{CHO} + ^{CH_3CO}_{CH_3CO} > O = C_6H_4 < ^{OH\ +\ CH_3COONa}_{CH=CHCOOH}$$

The resulting product, coumaric acid, loses water and forms coumarine, which thus becomes the anhydride of oxycinnamic acid :

$$C_6H_4 < ^{OH}_{CH=CHCOOH} = C_6H_4 < ^{O------}_{CH=CHCO\ \ |} + H_2O$$

Perkin[1] opposed this view at first; he could not admit that coumarine behaved like an anhydride. However, numerous investigations soon proved the correctness of Fittig's formula, and Perkin acknowledged himself mistaken. The matter rested until 1877, when Perkin[2] prepared cinnamic acid by the action of acetic anhydride and sodium acetate on benzaldehyde. The simplest formulation of this reaction, on the lines of Fittig's explanation, was the following:

$$\begin{matrix} C_6H_5CHO \\ C_6H_5CHO \end{matrix} + \begin{matrix} CH_3CO \\ CH_3CO \end{matrix} > O = \begin{matrix} C_6H_5CH=CHCOOH \\ C_6H_5CH=CHCOOH \end{matrix} + H_2O$$

But, unfortunately, it had been long since shown by Geuther[1] and Huebner[2] that benzaldehyde and acetic acid form not cinnamic acid, but benzylidene diacetate:

$$C_6H_5CHO + \begin{matrix} CH_3CO \\ CH_3CO \end{matrix} > O = C_6H_5CH < \begin{matrix} O-COCH_3 \\ O-COCH_3 \end{matrix}$$

Once more sodium acetate is found to play a mysterious part in this condensation. Perkin's[3] attempts to clear matters up served but to deepen the mystery. For on adding to the mixture of benzaldehyde and acetic anhydride sodium *butyrate* or *valerate* in lieu of the acetate, and heating to 180° as usual, he obtained cinnamic acid exclusively. If, however, sodium *propionate* and *propionic* anhydride, or sodium *butyrate* and *butyric* anhydride, were employed together, homologues of cinnamic acid were formed; viz. phenylcrotonic resp. phenylangelic acid. To these acids Perkin ascribed the formulæ:

$$C_6H_5CH=CH-CH_2-COOH, \quad \text{Phenylcrotonic acid}$$
$$C_6H_5CH=CH-CH_2-CH_2-COOH, \quad \text{Phenylangelic acid}$$

Succinic anhydride and sodium succinate reacted with benzaldehyde with evolution of carbon dioxide; a substance isomeric with phenylcrotonic acid

[1] *Ann. Chem.* (Liebig), 106, 251 (1858).
[2] *Ztschr. Chem.* 1867, 277.   [3] *l.c.*

resulted, which was called isophenylcrotonic acid. Perkin expressed this reaction thus:

$$C_6H_5CHO + \begin{array}{c} CH_2CO \\ | \\ CH_2CO \end{array}\!\!\!\!>\!O = \begin{array}{c} C_6H_5CH=C-COOH \\ | \\ CH_2COOH \end{array}$$

$$= \begin{array}{c} C_6H_5CH=C-COOH \\ | \\ CH_3 \end{array} + CO_2$$

His general conclusions were that in all cases the salt from which the anhydride is derived must be present, and that probably the carbon atom farthest removed from the carboxyl group is the one which condenses with the aldehyde. Of this latter idea, however, no experimental proof was offered.

The latter tentative hypothesis soon fell before a mass of contradictory evidence. Fittig[1] showed that unsaturated acids possess a double bond in a–β position to the carboxyl if they lose carbon dioxide when treated in succession with hydrobromic acid and an alkali. All of Perkin's acids gave this reaction, and therefore belonged in the category of $a\beta$ unsaturated compounds. Baeyer and Jackson[2] proved the following structure for phenylangelic acid:

$$\begin{array}{c} C_6H_5CH=C-COOH \\ | \\ CH_2CH_3 \end{array}$$

And finally, Conrad and Bischoff[3] demonstrated that Perkin's formulæ for phenylcrotonic and isophenyl-crotonic acids must be interchanged:

b_- [1] *Ann. Chem.* (Liebig), **195**, 169 (1879).
[2] *Ber. d. chem. Gesell.* **13**, 115 (1880).
[3] *Ann. Chem.* (Liebig), **204**, 183 (1880).

$$C_6H_5CH=C-COOH$$
$$| $$
$$CH_3$$

Phenylcrotonic acid

and $C_6H_5CH=CH-CH_2COOH$, Isophenylcrotonic acid

This much was certain, then, that in Perkin's synthesis it is the *alpha* carbon atom which reacts with the aldehyde. The part played by the salt of the acid, the presence of which Perkin had found to be an absolute necessity, was cleared up soon after by two incidental observations of Fittig and Jayne,[1] during the course of the investigation on unsaturated acids already referred to. In the first place, it was found that in the synthesis of isophenylcrotonic acid, succinic anhydride could be replaced by *acetic* anhydride with great advantage; also, that Perkin's working temperature, 180°, was much higher than necessary, 100° sufficing. And secondly, working under these improved conditions, they discovered that the first product of the reaction is a new substance, viz. phenylparaconic acid:

$$C_6H_5CH-CH-COOH$$
$$|\quad\quad|$$
$$|\quad\quad CH_2$$
$$|\quad\quad|$$
$$O-\!-\!-CO$$

This anhydride loses carbon dioxide on heating to 150°, and passes into *isophenylcrotonic acid:*

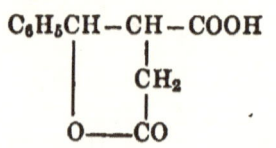

Three conclusions may be drawn from these unexpected results. Firstly, condensation occurs

[1] *Ann. Chem.* (Liebig), **216**, 100 (1883).

between the aldehyde and the *sodium salt;* the anhydride plays merely a secondary part.   Secondly,
the separation of water is preceded by an addition
process, phenylparaconic acid being the anhydride
or lactone of phenylitamalic (phenyloxypyrotartaric)
acid :

$$C_6H_5-CH(OH)-CH-COOH$$
$$|$$
$$CH_2COOH$$

Thirdly, the separation of carbon dioxide is also a
secondary process.

The first of these conclusions was in direct contradiction with the results obtained by Perkin ; as
mentioned above, he had found that benzaldehyde
and acetic anhydride always yielded cinnamic acid,
no matter whether sodium acetate, butyrate, or valerate was employed.   But experiments carried out
by Slocum,[1] at Fittig's suggestion, proved that Perkin's results were due entirely to the unnecessarily
high temperature he used; for if the temperature
be kept not much above 100°, acids were obtained
corresponding to the salt taken, irrespective of the
anhydride present.   Stuart,[2] another pupil of Fittig,
showed that in the presence of acetic anhydride the
salts of malonic and isosuccinic acids also condense
with benzaldehyde ; these acids are incapable of
forming anhydrides.

The correctness of the second conclusion was
rendered extremely probable by various cases in
which lactonic acids, intramolecular anhydrides of

[1] *Ber. d. Chem. Gesell.* **16**, 1436 (1883); *Ann. Chem.* (Liebig),
227, 55 (1885).

[2] *J. Chem. Soc.* **1883**, 403.

the original oxyacids, result from the reaction. Direct addition products of aldehydes with the sodium salts of *mono*basic acids have not yet been obtained. In one case, however, Fittig, together with Jayne[1] and Ott,[2] was able to isolate the butyryl derivative of such an addition product (the butyryl derivative resulting from the action of butyric anhydride on the hydroxyl group). The line of reasoning followed ran thus : If the reaction really consists in the union of aldehyde oxygen with two hydrogen atoms of the salt, with direct formation of an unsaturated acid (or its salt), then such acids as contain but *one* hydrogen atom attached to the *a*-carbon atom ought not to react with aldehydes. If, on the other hand, the process is simply one of addition, there is no reason why such an acid should not react. As a matter of fact, Fittig found that sodium isobutyrate, treated in the usual manner, yielded the butyryl compound of phenyloxypivalinic acid :

$$C_6H_5CHO + H-\underset{\underset{CH_3}{|}}{\overset{\overset{CH_3}{|}}{C}}-COONa \; = \; C_6H_5CH(OH)\underset{\underset{CH_3}{|}}{\overset{\overset{CH_3}{|}}{C}}-COONa$$

With these results, the history of Perkin's reaction is closed. In most respects, its mechanism has been cleared up, so that it constitutes one of the best-understood chapters of organic chemistry. However, its own history has shown how chary we must be of that stamp of self-approval, Q.E.D. Twice during the fifteen years here recorded did it seem as if all dubious points had been settled. Even to-day,

---

[1] *Ann. Chem.* (Liebig), **216**, 115 (1883).   [2] *Ibid.* **227**, 119 (1885).

with all the insight Fittig has given us, the function of the acid anhydride is as uncertain as before.[1] Who knows but that another revision will be made necessary when further results come home?

[1] Nef explains the purpose of acetic anhydride in this reaction as first forming benzylidene diacetate. *Ann. Chem.* (Liebig), **298**, 309 (1897). It is too early to assign a place in the history of Perkin's reaction to this view.

CHAPTER III

THE CONSTITUTION OF BENZENE

FARADAY, in 1825, isolated a liquid from coal gas, to which he gave the name "bicarburetted hydrogen." Nine years later, Mitscherlich found that the same substance was obtained by the dry distillation of benzoic acid with lime. These famous chemists little thought that their limpid oil would once lay claim to be the most important substance in organic chemistry, that it would give birth to untold thousands of compounds, that it would revolutionize science and technology.[1] The technical development of benzene and its derivatives employs over fifteen thousand workmen in Germany alone ; the commercial value of the products reaches tens of millions of dollars ; by far the greater portion of the research work done to-day is concerned with the same group of substances. Nearly all of this tremendous activity is due to a single idea, advanced in a masterly treatise by August Kekulé in the year 1865. Twenty-five years sufficed for the chemists of all nations to recognize the inestimable importance of the benzene theory, for in 1890[2] they came together at Berlin to do honor to the man who had created a new epoch in the science.

[1] Interesting details are given by H. Caro, *Ber. d. chem. Gesell.* **25**, Ref. 955 (1892).

[2] For a full account see *Ber. d. chem. Gesell.* **23**, 1265 (1890).

Early in the sixties, organic chemistry was just emerging from the type theory. The structural method of formulation was in the ascendant, and mainly through the efforts of Frankland, Williamson, Kekulé, Kolbe, and Butlerow, the genetic relations of the fatty or aliphatic series had been made clear. In contradistinction to these compounds stood a group of "aromatic" substances, not as yet very numerous, which in composition and behavior differed widely from the former, and which seemed to form a marked exception to the laws of valence and structure laid down for all carbon compounds. It was here that Kekulé surmounted the difficulty.[1]

"In attempting to give an account of the atomic constitution of the aromatic compounds, the following facts must be taken into consideration :

"(1) All aromatic compounds, even the simplest, contain more carbon, comparatively speaking, than analogous compounds of the fatty series.

" (2) There are many homologous substances in the aromatic series, just as in the fatty; *i.e.* substances whose difference in composition may be expressed by $n \cdot CH_2$.

" (3) The simplest aromatic compounds contain at least six atoms of carbon.

" (4) All products derived from the aromatic compounds present a certain family resemblance ; they all belong to the group of aromatic substances. A portion of the carbon is frequently eliminated during more energetic reactions, but the chief products contain at least six atoms of carbon (benzene, quinone, chloranil, phenol, oxyphenic[2] acid, picric acid, etc.). The de-

[1] *Bull. Soc. Chim.* **3**, 98 (1865); *Ann. Chem.* (Liebig), **137**, 129 (1866); *Lehrbuch*, II, 493.

[2] The old name for pyrocatechol.

composition ceases when these products are formed, except when complete destruction of the organic molecule takes place.

"These facts apparently justify the conclusion that all aromatic substances contain one and the same group of atoms, or, if you will, a common nucleus of six atoms of carbon. Within this nucleus the carbon atoms are to a certain extent in closer connection or conjunction. Other carbon atoms may then add themselves to this nucleus, in the same manner and according to the same laws that obtain in the fatty series.

"One must therefore first give an account of the atomic constitution of this nucleus. . . . "

Kekulé then proceeds to develop his now classic views on the constitution of benzene. He points out that the six carbon atoms are probably alternately connected by single and double linkings, to form a closed chain or ring. The remainder of the treatise is devoted to a detailed consideration of the consequences which follow from this conception. First, the constitution of many of the better known aromatic substances is given; in this section the principle is clearly laid down, that *each* carbon side-chain is converted into carboxyl upon oxidation — to this day one of our most important diagnostic reactions.

Kekulé's discussion of the possibilities of isomerism is extremely interesting. The most important case is where the substituents are located in the benzene ring itself. Two alternatives presented themselves : either all six of the hydrogen atoms are equivalent, or they are not. Each of these led to widely different conclusions.

I. *All the hydrogen atoms are chemically equivalent.* If so, only *one* monosubstituted benzene can exist,

and the benzene molecule is best represented by a
hexagon :

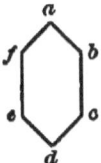

It can make no difference whether hydrogen atom
a or *b*, or any other, be replaced by, *e.g.*, bromine ;
the compounds are identical :

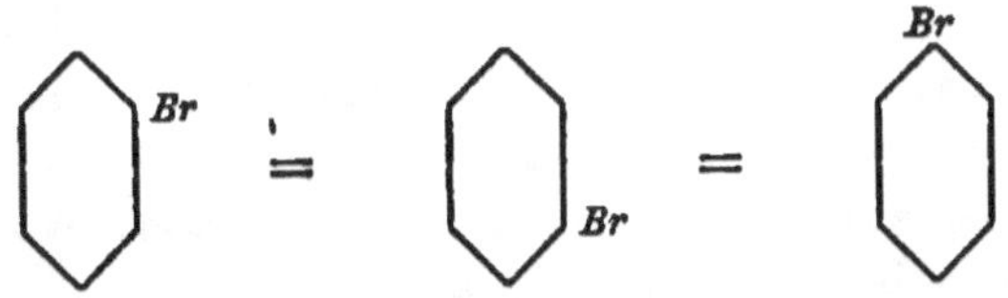

If *two* hydrogens are replaced, the theory requires
the existence of three, *and only three*, disubstituted
benzenes, according to whether the positions occupied
are *ab*, *ac*, or *ad* ; for position *ae* is manifestly identical with *ac*, and *af* with *ab*.

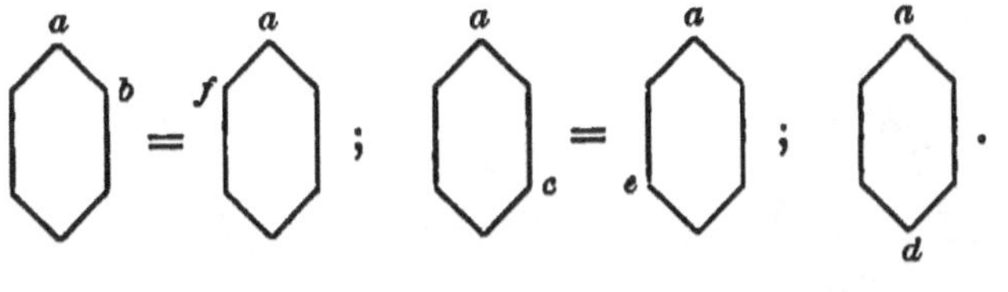

If three atoms of hydrogen are substituted by the
same radical, there must likewise be three isomeric
compounds :

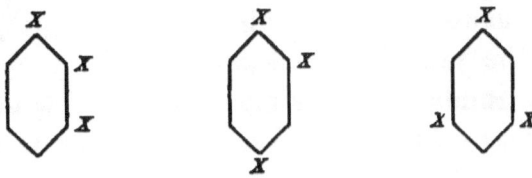

whereas if one of the three substituents be different from the other two, there are six possibilities :

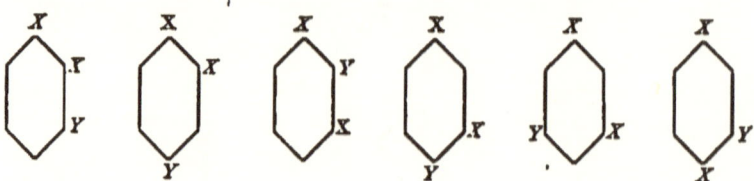

II. *The hydrogen atoms are not equivalent.*

" The six hydrogen atoms form three groups, each consisting of two carbon atoms united by two units of affinity each. The group, therefore, appears as a triangle, and we may further consider the atoms composing it so arranged that three atoms of hydrogen are inside the triangle, the other three outside. The six hydrogen atoms are then alternately unequal in value, and benzene may be represented by a triangle. Three of the six hydrogens are located at the angles — they are easily accessible; the other three are at the middle of the sides, so to speak in the interior of the molecule."

Kekulé has not expressed himself very clearly; but he means that *two* monosubstitution products of benzene must exist :

and a correspondingly greater number of di- and tri-derivatives.

Kekulé inclined to the first view, but confessed himself unable to prove his point. He proposed to attack the problem experimentally, by preparing as large a number as possible of benzene substitution products, by as many different methods as possible, and carefully examining them with regard to isomerism. By means of purely hypothetical assumptions, however, he made an attempt to determine the constitution of the di-brom- and brom-nitrobenzenes. These deductions were shown to be untenable shortly afterward, and can be omitted here.

The effect of this treatise of Kekulé's was electrical. In part or in their entirety, his views were adopted by almost every chemist working in the field of aromatic compounds. A vast number of experimental researches along the lines laid down in this brief guide, began to crowd the journals. The development was in three directions, along each of which were many ups and downs. The literature in question is enormous in extent ; in the limited scope of this book, account can be taken only of the work which directly contributed to the advance of the subject. The three lines mentioned are as follows :

(1) The equivalence of the six hydrogen atoms, and the possible number of isomeric substitution products.

(2) The genetic relations of aromatic compounds, with reference to the positions occupied by the substituents.

(3) The actual distribution of the six valence units which the *simple* hexagonal formula does not account for.

The investigation of these problems proceeded for the most part simultaneously, but for our purposes it will be best to treat them separately.

This epoch in the history of chemistry may fitly be compared to that following Dalton's announcement of the atomic theory. A very limited number of facts had formed the basis of a brilliant and comprehensive generalization. The expectancy of the scientific world was roused to the highest pitch, and with feverish haste and activity all rushed to the chemical Eldorado to participate in its promised wealth of knowledge and insight. A similar rush of scientific adventurers occurred three years ago, when Roentgen published his discovery of the X-rays.

I. THE RELATIONS OF THE HYDROGEN ATOMS

At the time when Kekulé advanced his theory, and expressed the opinion that only *one* form of monosubstituted benzene exists, there were two substances recorded which appeared flatly to contradict this assumption. The first of these was salylic acid, prepared by Kolbe by reduction of salicylic acid, and found by him to be isomeric with benzoic acid. Kekulé expressed his conviction that salylic acid was nothing other than impure benzoic acid; a view which was experimentally confirmed by Beilstein.[1] The other substance was a second modification of pentachlorbenzene, prepared by Jungfleisch. Kekulé was dubious as to the bona-fide existence of this compound, and trusted that in time the truth would be known. Ladenburg[2] (1874) proved that only one modification of pentachlorbenzene exists, the supposed isomere being a mixture of several substances. Thus, no fact is known which contradicts the assumed equivalence of the six hydrogen atoms in benzene.

[1] *Ann. Chem.* (Liebig), **132**, 151, 309 (1864).
[2] *Ann. Chem.* (Liebig), **172**, 331 (1874).

Negative arguments are not convincing or satisfactory. And so Ladenburg busied himself to find positive proof of Kekulé's proposition. He was able to show,[1] from the various formations of thymoquinone, that in benzene two pairs of hydrogen atoms are symmetrically situated with reference to one of the remaining atoms. This, historically the first proof, is of rather a complicated nature, and not easily followed ; as several very much clearer demonstrations of the theorem have been given, it can be safely omitted in detail.

The existence of *one* pair of symmetrical hydrogen atoms had been shown considerably earlier by Huebner and Petermann[2] (1869). Brombenzoic acid yields *two* nitrobrombenzoic acids. If these are reduced to the amido acids (which are still different), and then by further reduction the bromine eliminated, the resulting amidobenzoic acids are identical. Now in the nitrobrom- (and amido-) brombenzoic acids, the difference in properties must be caused by differences in position of the nitro (and amido) groups. As, however, by removal of the bromine atom only *one* amidobenzoic acid is obtained from two substances, the amido (and nitro) groups in these latter must be symmetrically situated with respect to the carboxyl group, as otherwise the final products could not be identical.

The best proof by far of the existence of two symmetrical pairs of atoms in the benzene ring was given

---

[1] *l.c.*, p. 344, *Theorie der arom. Verbindungen*, p. 14 (1874); Ladenburg and Engelbrecht, *Ber. d. chem. Gesell.* 10, 1218 (1878).

[2] *Ann. Chem.* (Liebig), 149, 130 (1869) ; cf. Ladenburg, *Ber. d. chem. Gesell.* 2, 140 (1869), for this explanation.

by E. Wroblewsky,[1] in 1878. Wroblewsky prepared all five possible brom-benzoic acids, in order to see how many are identical. The chart on page 31 will serve to illustrate the reactions which he employed for this purpose. His starting-point was a brom-toluidine (I). By eliminating the amido group by means of the diazo reaction, a brom-toluene (II) results, which upon oxidation gives meta-brombenzoic acid (III). This concludes one series with a well-known compound. Starting again with brom-toluidine (I), nitration (IV) with subsequent elimination of the amido group gives a nitro-brom-toluene (V), which is reduced to *another* brom-amido-toluene (VI). Further reduction removes the bromine (VII) ; the resulting meta-toluidine is diazotized, the diazo group replaced by bromine, and we have a brom-toluene (VIII) identical with II, which is oxidized to a brom-benzoic acid (IX) identical with III. Position 1–3 is therefore equivalent to position 1–5, and hydrogen atoms 3 and 5 form a symmetrical pair with respect to hydrogen atom 1. Wroblewsky's third series again commences with compound I. Preparing IV as above, the amido group is exchanged for an iodine atom by means of the diazo reaction (X). Reduction of this new substance gives a brom-iodo-toluidine (XI), whose amido group is likewise exchanged for a bromine atom (XII). We now have a toluene with three substituents, leaving only two vacant places in the ring open for attack. Nitration forces a nitro group into one of these two

---

[1] *Ann. Chem.* (Liebig), **192**, 196 (1878).

places (it makes no difference which one we assume to be thus occupied). The dibrom-iodo-nitro-toluene (XIII) thus obtained is reduced to the corresponding toluidine (XIV), further reduction removing all the halogen atoms and leaving ortho-toluidine (XV). This is now converted into brom-toluene (diazo-reaction) (XVI), and finally oxidized to ortho-brombenzoic acid (XVII). Thus we have established the formula of a third brom-benzoic acid. The fourth series begins with compound XIV. By means of the diazo reaction, the amido group is replaced by iodine (XVIII). The resulting dibrom-diiodotoluene possesses but one replaceable hydrogen atom, *the one which in* XIII *was not attacked by nitric acid.* This last hydrogen is exchanged for the nitro group (XIX), reduction yields an amine (XX), and energetic treatment with nascent hydrogen removes all of the halogen atoms. The product is ortho-toluidine (XXI), identical with (XV). The bromtoluene (XXII) and brom-benzoic acid (XXIII) which it gives, are of course identical with the substances XVI and XVII, and thus our fourth brombenzoic acid is one and the same with the third, thereby establishing the existence of a *second* symmetrical pair of hydrogen atoms. As to the fifth brom-benzoic acid, it is readily prepared from paratoluidine by reactions similar to the ones just discussed, and is different from either of the other two (or four).

This important demonstration thus clearly proves the existence of the symmetry postulated by Kekulé, and further shows that three, *and only three*, disubstitution products of benzene can exist. A year later, it was supplemented by Huebner,[1] who completed his previous work by illustrating a second pair of correlated hydrogen atoms. Salicylic acid

gives two nitro derivatives. These nitro-hydroxy-benzoic acids are converted into nitro-amido-benzoic

[1] *Ann. Chem.* (Liebig), **195**, 1 (1879).

acids upon boiling with ammonia. These two acids, when the amido group is removed (diazo reaction), give one and the same nitro-benzoic acid. Therefore they contain nitro groups symmetrically arranged with reference to the carboxyl group. This nitro acid reduces to an amido-benzoic acid different from the one Huebner obtained previously, and the two nitro groups thus correspond to a symmetrical pair of hydrogens differing from the previous pair.

Upon the basis of what is thus proved, viz. that only three di-substituted benzenes can exist,

$$ab = af, \ ac = ae, \ \text{and} \ ad,$$

Ladenburg[1] was able to show the *equivalence of all six hydrogen atoms in benzene.* Phenol passes into brom-benzene when treated with phosphorus penta-bromide ; and brom-benzene is converted into benzoic acid when treated with carbon dioxide and sodium. We thus pass from phenol to benzoic acid, and if the idea of definite positions in the ring has any meaning at all, the hydroxyl of the one must occupy the same place as the carboxyl of the other. We may call this position a. Now three oxy-benzoic acids are known. These may all be reduced to ordinary benzoic acid, and therefore they contain their carboxyl group in position a. The hydroxyl groups must be in three different positions, as the acids are different ; these may be designated as b, c, and d. When the oxybenzoic acids are heated with lime, carbon dioxide is split off, and phenols result which must contain their

---

[1] *Ber. d. chem. Gesell.* 7, 1684 (1874).

hydroxyl respectively in positions **b**, **c**, and **d**.
These phenols were found to be *identical with
ordinary phenol,* which contains hydroxyl in **a**.
Therefore it makes no difference whether a monosubstituted benzene contains the substituent in **a**,
b, **c**, or **d**; the compounds are identical, and these
positions must be absolutely equivalent. Among
the three hydrogens **b**, **c**, and **d**, no two are symmetrical towards **a**, for the three oxybenzoic acids
could not then be different from one another. The
remaining atoms **e** and **f** must be situated with
regard to **a** just as *two* of the atoms **b**, **c**, and **d**.
The oxybenzoic acids

$$C_6H_4(CO_2^aH)(O\overset{e}{H}) \quad \text{and} \quad C_6H_4(CO_2^aH)(O\overset{f}{H})$$

must be identical with two of the known oxybenzoic
acids,[1] and the phenols resulting from them

$$C_6H_5(O\overset{e}{H}) \quad \text{and} \quad C_6H_5(O\overset{f}{H})$$

identical with ordinary phenol. Therefore all six
positions in the benzene ring are equivalent; no
matter which of them is substituted by a radical,
the resulting compound is the same.

II. THE GENETIC RELATIONS OF AROMATIC COMPOUNDS

Kekulé made an attempt to determine the location
of substituents by deduction from unfounded general
principles. It was found later that no general principles obtained here, that the problem was essentially

---

[1] This follows from Wroblewsky's work.

experimental in its nature. Koerner, a pupil of
Kekulé's, proposed to distinguish the three series of
isomeric compounds by the prefixes still in use, *ortho*,
para, and *meta*. At first there was great confusion,
until Kekulé suggested that all compounds which
were shown to be related to a standard ortho compound should be called *ortho*, etc. These group
names were in use for a long while before it was
known to what actual differences in position these
designations corresponded. Later it was found that
most of the substances termed *para* contained their
substituents in the 1-4 position, and therefore this
prefix was adopted for such substances only. The
same may be said of the *ortho* and *meta* compounds.

It required over ten years of incessant labor on
the part of Koerner, V. Meyer, Griess, Ladenburg,
and many others to finally establish fixed standards
of reference and fit all aromatic compounds to them.
The first step was taken by Baeyer[1] (1868), who
remarked that the formation of mesitylene from acetone was best explained by assuming a symmetrical
structure for the former; and as Fittig had shown
mesitylene to be trimethyl benzene, its formula could
be written:

A few years later (1869) Graebe[2] found that naphthalene was to be represented by the following formula:

[1] *Ann. Chem.* (Liebig), **140**, 306 (1866).
[2] *Ibid.* **149**, 1 (1869).

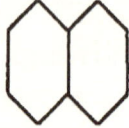

As naphthalene on oxidation furnished phthalic acid, the constitution of the latter was thus fixed as 1–2 (ortho):

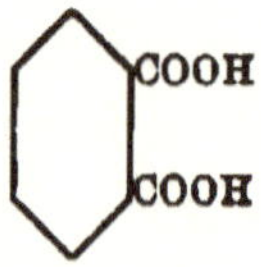

the formula still given to it. At about the same time Ladenburg[1] pointed out that para compounds must have the position 1–4; this left 1–3 for the meta compounds.

All this seemed simple enough; substances which can be converted into phthalic acid belong to the ortho series, containing their substituents in 1–2 position; etc. The trouble lay in the fact that the reactions selected to show genetic relations were frequently absolutely unreliable. Considerable differences of opinion obtained until a sufficiently large number of facts had been amassed to weed out the disturbing irregularities. For the sake of the interest attaching to these erroneous views, a short list of them will be given.

Graebe[2] considered hydroquinone and metaoxy-benzoic acid to be ortho compounds, in the same series with phthalic acid: salicylic acid and pyrocatechol as meta, resorcinol as para. Huebner and Petermann[3] regarded ordinary brombenzoic acid (meta) as

[1] *Ber. d. chem. Gesell.* **2**, 140 (1869).   [2] *l.c.*   [3] *l.c.*

belonging in the same category with salicylic acid, which was therefore also meta; as anthranilic acid is directly convertible into salicylic acid, it also was a meta compound.   As late as 1871, V. Meyer[1] adopted the classification of the dioxybenzenes given by Graebe; quinone, resulting from hydroquinone by simple oxidation, must likewise be ortho.   In the same paper, dinitrobenzene (meta) is regarded as 1-4, volatile nitrophenol (ortho) as 1-3, non-volatile nitrophenol (para) as 1-2.   Salicylic acid is shown to be an ortho compound.   The following year Salkowsky and Zincke[2] independently showed that the volatile nitrophenol belongs to the ortho series.   In 1873 Huebner and Schneider[3] agreed with this view, and suggested that the non-volatile compound is 1-4.   These authors are still uncertain as to the exact relations of the dioxybenzenes.   During the same year Wroblewsky[4] clearly showed the relations between a large number of disubstituted products.   V. Meyer[5] (1874) says that dinitro-benzene and resorcine are members of the same series (according to Koerner), and that resorcine is un-doubtedly connected with terephthalic acid; therefore dinitrobenzene and resorcine must be para compounds.   Quinone, he says, is now universally conceded to be 1-3.[6]   Pyrocatechol is shown to be an ortho derivative.

---

[1] *Ann. Chem.* (Liebig), **159**, 21 (1871).

[2] *Ber. d. chem. Gesell.* **5**, 114, 874 (1872); *ibid.* **6**, 123, 140 (1873).

[3] *Ann. Chem.* (Liebig), **167**, 114 (1873).

[4] *Ann. Chem.* (Liebig), **168**, 147 (1873).

[5] *Ann. Chem.* (Liebig), **171**, 57 (1874).

[6] Cf. above.

These few examples will suffice to show in how unsettled a condition the problem remained for a long period. The chief cause thereof was that rearrangements took place during the reactions relied upon to give a clear genealogy, so that the substituents *after* reaction were no longer in the same relative positions as before. For instance, all three phenol sulfonic acids[1] were found to give the same dioxybenzene on fusion with alkalies, viz. resorcinol. The action of potassium cyanide on bromnitro compounds, a reaction largely relied upon to furnish substituted benzoic acids, was shown to be accompanied by similar rearrangements,[2] so that the material gathered in this way was quite unavailable for the purpose in question.

An exact and thoroughly reliable basis for the determination of the constitution of aromatic compounds was furnished by Koerner,[3] Griess,[4] and Salkowski,[5] independently of one another. This consisted in showing what the relations are, and must be, which obtain between di- and tri-substituted benzenes. Suppose that the di-derivatives of the formula $C_6H_4X_2$ are converted into tri-derivatives $C_6H_3X_2Y$; where Y may be identical with X, or different from it. It is clear that an ortho compound will give *two* tri-derivatives :

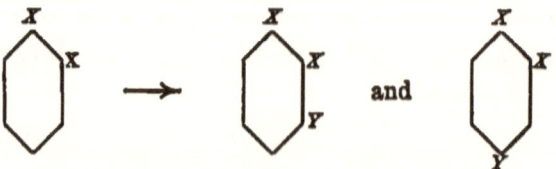

[1] Cf. Kekulé, *Ber. d. chem. Gesell.* 2, 330 (1869).
[2] V. Richter, *Ber. d. chem. Gesell.* 8, 1418 (1875).
[3] *Gazz. Chim. ital.* 4, 443 (1875).
[4] *Ber. d. chem. Gesell.* 7, 1226 (1874).
[5] *Ann. Chem.* (Liebig), 173, 66 (1874).

Similarly, a meta compound will give *three*:

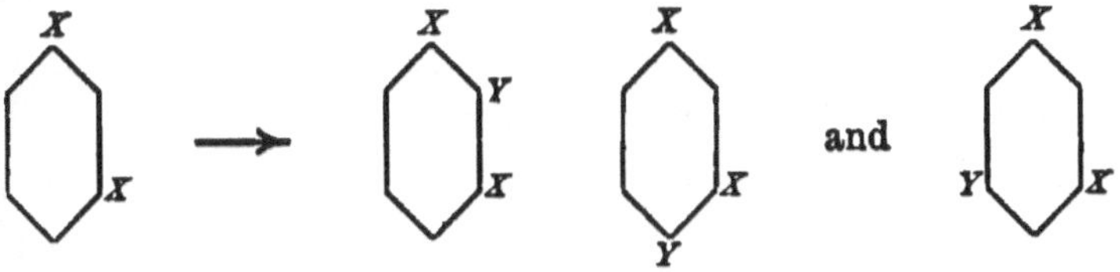

The para compounds can yield only *one*:

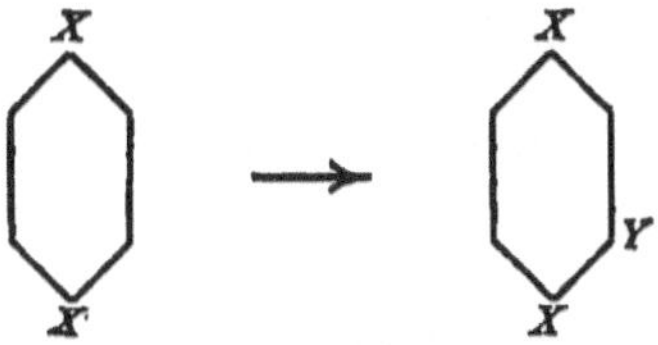

Koerner used the dibrombenzenes to illustrate this principle. He found that the dibrombenzene boiling at 220° gave three tribrom- and three nitrodibrombenzenes, and must therefore be the meta compound. Another dibrombenzene (melting at −1° and boiling at 224°) gave two nitro derivatives and two tribrom compounds, and is thus the ortho member. The third dibrombenzene must therefore be the para compound; it in fact furnished only one nitro product and one tribrombenzene. Griess and Salkowski employed the relations between the six isomeric diamidobenzoic acids and the three diamidobenzenes. On splitting off carbon dioxide from the acids, *three* gave a phenylendiamine melting at 62°, *two* a diamine melting at 102°, *one* a diamine melting at 140°. The first-mentioned diamidobenzene is therefore a meta, the second an ortho, the third a para compound.

A number of other investigations on this problem of the constitution of benzene derivatives have been published, of which only one more need be considered here. This is Ladenburg's[1] proof of the equivalence of the three non-substituted hydrogen atoms in mesitylene (thus showing its symmetrical structure). On nitrating the hydrocarbon a dinitro product is obtained ; the nitro groups may be considered in positions a and β ; by reduction of one of the nitro groups (let us say β), a nitromesidine is obtained :

$$C_6(CH_3)_3(\overset{a}{N}O_2)(\overset{\beta}{N}H_2)(\overset{\gamma}{H}).$$

If the amino group be acetylated, a further nitro group may be introduced (by direct nitration):

$$C_6(CH_3)_3(\overset{a}{N}O_2)(NH.\overset{\beta}{C}_2H_3O)(\overset{\gamma}{N}O_2).$$

When this dinitroacetmesidine has its $-NH.C_2H_3O$ group replaced by hydrogen (by first saponifying the acetyl group and then eliminating the amino group by means of the diazo reaction), a dinitromesitylene is obtained :

$$C_6(CH_3)_3(\overset{a}{N}O_2)(\overset{\beta}{H})(\overset{\gamma}{N}O_2),$$

which is identical with the original dinitromesitylene :

$$C_6(CH_3)_3(\overset{a}{N}O_2)(\overset{\beta}{N}O_2)(\overset{\gamma}{H}).$$

Therefore hydrogen atoms β and γ are equivalent. If, further, in the above-mentioned nitromesidine,

$$C_6(CH_3)_3(\overset{a}{N}O_2)(\overset{\beta}{N}H_2)(\overset{\gamma}{H}),$$

the amido group be eliminated (by means of the diazo reaction), the nitro group then reduced to

[1] *Ann. Chem.* (Liebig), **179**, 174 (1875).

amino and this acetylated, there results an acetmesidine :

$$C_6(CH_3)_3(NH.\overset{a}{C}_2H_3O)(\overset{\beta}{H})(\overset{\gamma}{H}).$$

When this is nitrated, the entering nitro group must be either in position β or γ :

$$C_6(CH_3)_3(NH.\overset{a}{C}_2H_3O)(\overset{\beta}{NO}_2)(\overset{\gamma}{H})$$

or $$C_6(CH_3)_3(NH.\overset{a}{C}_2H_3O)(\overset{\beta}{H})(N\overset{\gamma}{O}_2).$$

But these formulæ are identical, since, as just shown, positions β and γ are equivalent. If from this nitroacetmesidine the acetyl group be removed, there results a nitromesidine :

$$C_6(CH_3)_3(\overset{a}{NH}_2)(\overset{\beta}{NO}_2)(\overset{\gamma}{H}),$$

which proves to be identical with the nitromesidine taken as a starting-point :

$$C_6(CH_3)_3(\overset{a}{NO}_2)(\overset{\beta}{NH}_2)(\overset{\gamma}{H}).$$

Therefore atom a is likewise equivalent with atom β, and mesitylene has the symmetrical structure :

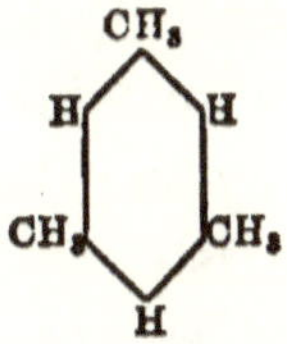

All subsequent work has been based on the rigid foundation furnished by these investigations. The problem of determining the structure of benzene compounds of course comes up with every new substance discovered. Many more complicated aromatic compounds still await their complete identification, which can only be a question of time and diligence.

III. THE DISTRIBUTION OF VALENCES IN THE BENZENE RING

We have seen that the comparatively short period of ten years sufficed abundantly to confirm the views of Kekulé as regards symmetry of the benzene ring, and the consequences resulting therefrom. Not so with the actual structure proposed by him at the same time. Kekulé's original formula, it will be remembered, was based on the assumption that the six carbon atoms were united by single and double linkings alternately:

It is not to be supposed that at that time Kekulé considered his formula to be a complete and thorough diagrammatic representation of all the properties of the benzene complex. The formula was merely the *simplest* expression thereof, under existing circumstances. Until the more pressing question of symmetry was definitely settled, that of structural detail was of a nature more suitable to interest debating societies than to occupy the thoughts and direct the efforts of earnest scientists.

However, it did not escape the watchful eyes of his contemporaries that Kekulé's own formula *precluded* some of the consequences he deduced from it: it admits of two isomeric ortho derivatives:

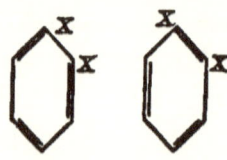

Kekulé was therefore compelled to issue an additional hypothesis[1] in aid of his main theory (1872). This hypothesis was as follows : If we suppose the atoms to be in continual motion, striking against one another as they vibrate, then the difference between a single and a double linking among the carbon atoms may be easily imagined to lie in that the former consists of one contact per unit of time, the latter of two per unit. A given carbon atom in the benzene ring would therefore be struck twice by, say, its right-hand neighbor, during the period in which its left-hand neighbor bounded against it once. Granted the plausibility of this assumption, there is no reason why these vibrations may not pass through a cycle of changes such that contact between two atoms occurs alternately once and twice in whatever period of time these motions require.[2] The double linking in this case would be alternately on the right and on the left side of the carbon atom, and as the oscillations are likely to be extremely rapid, to all practical intents and purposes position b would be identical with position f :

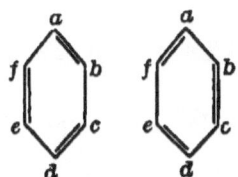

As the equivalence of the six hydrogen atoms is a proved fact, independent of any theory about the

[1] *Ann. Chem.* (Liebig), **162**, 77 (1872).
[2] Mechanically, this is rather difficult to conceive of. Cf. Michaelis, *Ber. d. chem. Gesell.* **5**, 463 (1872).

structure of the ring, this hypothesis of the "oscillating double linking" has become an integral part of Kekulé's theory.   Without it, the symmetry could not be accounted for.   However, during the years that elapsed between the first publication of the benzene theory and its emendation as just given, several other formulæ were proposed to fill the gap. Three of these had or have serious claims for consideration; they will be discussed in detail below. Others, such as the one given by Dewar (1866):

are lacking in symmetry, and therefore are not to be entertained for a single moment.

The three formulæ which have equal claims with Kekulé's, when considered with regard to the symmetry of the ring, are known respectively as the *prism*, the *diagonal*, and the *centric* formulæ.   They will be reviewed in the order named.

I. *The Prism Formula*

The prism formula was first seriously advocated by Ladenburg (1869).   This formula distributes the six extra valences as follows:

This diagram is the plane projection of a prism :

It may be seen at a glance that the prism is perfectly symmetrical ; there can be only one form of
mono-substitution product. The formula also accounts satisfactorily for the observed phenomena of
di-substituted benzenes : three, and three only, can
exist.   But here arises a peculiarity of the prism
formula ; if we remember how Koerner defined the
ortho, meta, and para positions (p. 37), and if we
further adhere to the established custom of number-
ing the ortho position 1–2, etc., then the carbon
atoms of the prism must *not be numbered in regular
order*, thus :

but as follows :

And for this reason : the para position is that which
furnishes but one tri-derivative. This condition is
satisfied by the vertical edges of the prism ; positions 5, 3, 2, and 6 all are symmetrical with respect
to 1–4.   The carbon atoms at the intersections of
the sides of the triangles stand in meta position
toward each other ; for of the four possible com-

binations containing, say, the atoms 1 and 3, viz. 1.3.5, 1.3.2, 1.3.4, and 1.3.6, only the last two are identical; the meta position corresponds to three tri-derivatives. The ortho atoms lie in different triangles, for position 1–2 yields two tri-derivatives.

Ladenburg's prism formula thus leads to a different conception of the relative positions of the atoms in benzene; for according to Kekulé's formula the ortho atoms are directly united, whereas Ladenburg places them apart. Ladenburg's para atoms are directly coupled, Kekulé's are not; etc. This separation of the ortho atoms was felt by Ladenburg[1] himself as an objection to the formula, though he continued its advocacy.

The prism formula, for the reason that it completely explains all the facts observed in connection with isomerism, is thus distinctly superior to Kekulé's, which requires the additional hypothesis of an oscillating double linking. Yet it is a strange fact that at no time did the prism formula supersede the older one. None of the investigators mentioned in the earlier part of this chapter, aside from Ladenburg himself, availed themselves of it. In spite of Ladenburg's superior logic it remained almost a scientific curiosity. This was partly caused by the spatial nature of the prism formula. Stereo-chemical considerations were regarded with extreme distrust at the time in question, and chemists carefully avoided them. The chief cause, however, lay in the unwieldy formulæ to which the prism led when ap-

---

[1] *Ann. Chem.* (Liebig), **179**, 174 (1875).

plied to the so-called "condensed" benzene rings;
e.g. naphthalene :

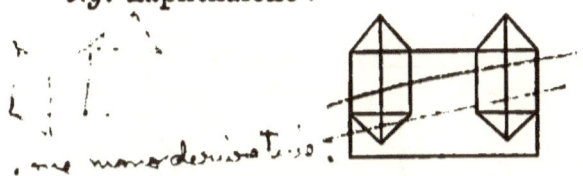

Such a structure is certainly not simple.

But interesting as the prism formula is, and excellent though the arguments in its favor, it can
no longer figure as a possibility.   In 1886, A. von
Baeyer began a series of experimental investigations
on the distribution of the valences in benzene, the
first result of which was to show the untenability of
the prism formula.   Baeyer studied the reduction
products of benzene derivatives, with the idea in
mind that the best way to get at the six valences
was to put some of them out of the road.   Now the
prism formula and the Kekulé formula lead to widely
different expectations when we loosen the *valences
not directly occupied in holding the ring together.*   In
Kekulé's formula no change in *relative* positions can
take place when we pass from a normal to a hexa-
hydrated derivative ; none of the linkings that determine the ortho or meta or para position are
loosened :

Not so with the prism ; on reducing it to the hexa-
hydro derivative, only one of the three para positions

remains intact.   This is readily seen from the following formulæ :

$$\text{formulæ diagram}$$

Of the three sets of para positions, viz. 1–4, 2–5,
3–6, but one remains a true para position (1–4); the
other two have assumed ortho relations.   If, therefore, a benzene compound containing two or three
sets of para substituents still contains two or three
para sets after being reduced, the prism formula cannot be correct.   Baeyer's experiments[1] proved this
to be the case.   Dioxyterephthalic ester has the following structure :

$$\text{structure diagram with } CO_2C_2H_5 \text{, OH, OH, } CO_2C_2H_5$$

because it readily yields hydroquinone.   On reduction with sodium amalgam it passes into succinylosuccinic ester.   This compound results from the
action of sodium on succinic ester, just as acetoacetic ester is formed from acetic ester ; and its
method of formation leads to the formula :

$$\text{structure diagram with } H\ CO.OC_2H_5 \text{, } H, OH, HO, H, H\ CO.OC_2H_5$$

[1] *Ber. d. chem. Gesell.* **19,** 1797 (1888).

Now in succinylosuccinic ester the hydroxyl groups are in para position to each other, likewise the carbethoxyl groups. The same is true of dioxyterephthalic ester. *Two* sets of para positions have thus remained unaltered, a result impossible under the prism formula.

The behavior of phthalic acid[1] upon reduction furnishes a similar argument against the prism. According to this latter, ortho compounds should be converted into meta :

But hexahydrophthalic acid behaves exactly like dimethylsuccinic acid, yielding an anhydride :

Hexahydroisophthalic acid,[2] on the contrary, which according to Ladenburg's theory should be an ortho compound :

[1] *Ann. Chem.* (Liebig), **258**, 145 (1890).
[2] Baeyer and Villiger, *Ann. Chem.* (Liebig), **276**, 259 (1893).

does not give an anhydride. The reduction of
phthalic and isophthalic acids therefore does not
alter the relative positions of the substituents. After
these results, the prism formula has not a leg left to
stand upon.[1]

II. *The Diagonal Formula*

A. Claus,[2] in a work now difficultly accessible, first
suggested that the constitution of benzene might
be better expressed by the following formula :

As originally proposed, this structure conflicted with
the proved existence of three di-substitution products ; for each carbon atom is directly connected
with *three* others :

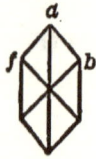   

Compounds **ab**, **af**, and **ad** ought therefore to be
identical — in contradiction to experience.

To remedy this defect in his formula, Claus[3]
announced that the " para-bonds " were not like ordi-

[1] Ladenburg admits the weighty evidence against his formula,
but contends that *some* benzene derivatives are so constructed.
Ber. d. Chem. Gesell. **20**, 62 (1887).

[2] *Theoretische Betrachtungen* (Freiburg, 1867), p. 207.

[3] Cf. *Ber. d. chem. Gesell.* **15**, 1407 (1882); **20**, 1423 (1887).

nary ones ; first of all, they connected atoms that were further apart than is ordinarily the case ; and secondly, they are much more easily ruptured.   The assumption of the influence of *distance* on affinity is of course purely hypothetical ; the length of line we *draw on paper* to represent the connection of atoms can have no possible effect upon the nature of that union.   But the assumption that these para-bonds were *somehow* different from ordinary bonds, was a point clearly capable of experimental investigation.   This investigation was carried out by Baeyer, in the series already referred to.   As Baeyer's papers form extremely complicated reading matter, on account of the fact that their author himself changed his point of view several times during the course of research, I can do no better than to give literally a portion of Baeyer's address [1] on the occasion of the Kekulé celebration, in which he sums up the evidence.

" *If this assumption (as to the difference of para-from ordinary bonds) is correct, it is to be expected that after rupturing one bond the presence of the other two could be shown.   Experiment proved, however, that only double linkings remained, and so it was concluded that benzene contained no para-bonds.   To this Claus objected that the two para-bonds might rearrange themselves to form double linkings :*

[1] *Berichte*, **23**, 1277 (1890).

"At the first glance it would seem easy to decide this question experimentally by reduction of terephthalic acid. Provided Claus is correct, the dihydro-acid first formed must have the formula :

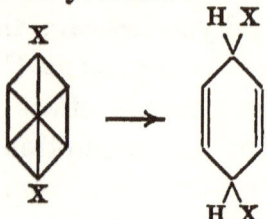

Kekulé's formula would lead to an acid of different constitution :

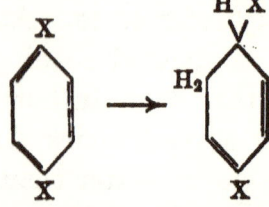

It was found that upon extremely careful reduction there resulted the compound indicated by Claus's view. But at the same time observations were made which showed that the very same result might be obtained on the basis of Kekulé's formula.

" The dihydro-acid :

upon reduction furnishes a tetrahydro-acid of the following constitution :

Here, too, the reduction has proceeded as if an easily loosened para-bond were present. In order to remove all doubt in this respect, I have, together with Dr. Rupe, reduced muconic acid, and have arrived at the very identical result :

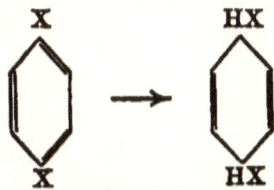

Muconic acid yields a hydromuconic acid of corresponding type, and it is thus shown that the addition of hydrogen in the para position by no means requires the presence of a para linking in the ring. The reduction of terephthalic acid may thus be explained by Kekulé's formula :

Baeyer thus arrives at the conclusion that no para-bonds can be found *experimentally* in the benzene ring.

III. *The Centric Formula*

Properly speaking, this is no separate formula, but merely a modification of Claus's. It was first proposed by Armstrong,[1] almost immediately thereafter by Baeyer,[2] and finally adopted by Claus in

---

[1] *J. Chem Soc.* **1887**, 264, 582.

[2] *Ann. Chem.* (Liebig), **245**, 118 (1888).

this modified form. It supposes that the extra six valences do not directly unite, but are attracted toward the centre and neutralize each other all of a heap, so to speak :

They are held in this position by delicately balanced forces, so that if anything should happen to one of them, the rest rearrange themselves according to the conditions of greatest stability.

There is no rigid proof of the centric formula. It may be said that there is no proof at all of it. In its nature it is incapable of proof, for the premise that the valences are held in position by forces that admit of no disturbance except by complete alteration, precludes experimental study. The formula is merely the best picture we have of the complete symmetry of the ring, of its stability toward reagents, and of its difference from the other, better-known forms of union between carbon atoms. *By definition* it avoids these difficulties. Benzene contains no para-bonds that we can lay hold of; if it contains double linkings, these are certainly not like the ordinary ones of the fatty series, and require a special hypothesis, both for their lack of additive powers and for the absence of isomeric ortho substitution products. Under such conditions it is much better to use an entirely characteristic formula than to adhere to one which is misleading except under interpretations of an uncertain nature.

Certain facts have led Baeyer to believe that there is no *one* benzene formula — that both the centric and the Kekulé formulæ are correct; in other words, that some benzene derivatives possess the one, some the other structure. In later papers than the one quoted, he advocates[1] a cross between the original and the modified Claus formula for phthalic acid; theoretical considerations would show that there is perhaps in this case a closer connection between the para atoms than is given by the centric formula. On the other hand, phloroglucine[2] behaves as if it really contained three double linkings. With hydroxylamine it furnishes an oxime, with phenyl-hydrazine a hydrazide; in other respects it is undoubtedly a trioxybenzene. The following formulæ explain the reactions mentioned :

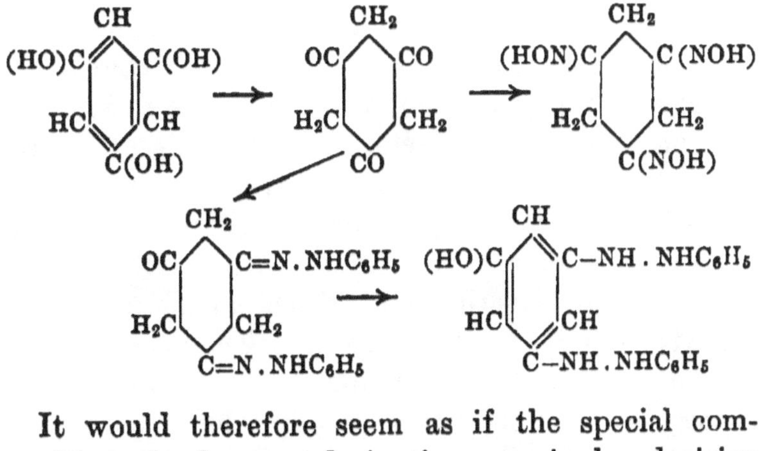

It would therefore seem as if the special composition of a benzene derivative exercised a decisive influence upon its final constitution : a result in thorough agreement with our experiences in other

[1] *Ann. Chem.* (Liebig), **269**, 172, 187 (1892).
[2] *Ber. d. chem. Gesell.* **19**, 159 (1886); **24**, 2687 (1891).

fields (cf. Chap. IV.). The "benzene problem" resolves itself into a series of special problems, from
the solution of which we may expect much light
to be shed upon the problem as a whole. Baeyer's
investigations on the peculiar nature of phthalic acid
have not been extended since the early part of 1892,
which is greatly to be regretted. Many examples of
the phloroglucine type have been observed, but for
the present they have added nothing to the benzene
theory.

A fairly complete account has now been given of
the work carried out with the special object in view
of determining the structure of benzene and its derivatives. Many other investigations have shed
sidelights upon the question, and hardly a single investigator has failed to seize his opportunity. The
most prominent of these, more from the discussions
they have raised than from actual gain to our knowledge of the problem, have been the researches upon
the so-called physical constants. As is well known,
both the molecular refraction and the molecular heat
of combustion are subject to "constitutive influ-
ences"; *i.e.* the values of these constants differ for
isomeric substances, the difference depending upon
differences in constitution. Chief among the constitutive details which must be taken into account
is the nature of the linking between the carbon
atoms, each double and each triple linking exercis-
ing a marked and approximately constant influence.
What could be simpler, then, than directly to observe
the refractive and thermochemical constants of benzene, and by comparison with the computed values
ascertain the constitution of this devious hydro-

carbon ?   Upon this as a basis, Thomsen[1] and Stohmann[2] came to the conclusion that the heat of combustion of benzene is incompatible with the presence
of double linkings, whereas Brühl[3] insisted that the
refractive constant proves the existence of three
double bonds in the very same substance.   This ludi-
crous contradiction is probably a mere coincidence ;
in fact, by taking another point of view, Brühl[4] has
endeavored to show that Stohmann's figures mean
the opposite of their author's conclusion.   Here lies
the chief difficulty about such determinations of constitution ; they depend largely upon one's point of
view, rather than upon fixed and definite standards.
We must remember that constants derived from
observations in the fatty series cannot safely be employed to prove or disprove the presence of group-
ings in the aromatic series.   If benzene contains
"double linkings," these must certainly be different
from ordinary fatty double bonds; else we should
not be put to such trouble to find them.   Whether
or not we employ the same mechanical device to
picture these two widely divergent somethings we
call double bonds, must not obscure the issue.   The
refractometric and thermochemical "constants" have
been found to be so extremely sensitive to slight constitutional changes, that the peculiarities of the benzene structure must have effects far beyond our
simple guessing powers to estimate.

Other attempts have been made to ascertain the

---

[1] *Ber. d. chem. Gesell.* **13**, 1808 (1880).
[2] *J. prakt. Chem.* [2], **48**, 453 (1893).
[3] *Ann. Chem.* (Liebig), **200**, 139 (1880).
[4] *Ber. d. chem. Gesell.* **27**, 1065 (1894).

secret of the constitution of benzene by indirect means.[1] They depend upon more or less uncertain analogies, and differ as widely in their conclusions as do the "physical" methods. Their main interest lies in this diversity of result rather than in their contribution to the benzene problem, and we may therefore omit their consideration here.

The new doctrines recently introduced into organic chemistry, which deal with the spatial arrangement of the atoms, have met with but moderate success in their application to the benzene ring. Certain forms of isomerism, it is true, may be easily explained by their aid. Thus Baeyer[2] was able to account for the existence of two hexahydro-terephthalic acids as follows :

$$\begin{array}{cc} H_2 \quad H_2 & H_2 \quad H_2 \\ \underset{COOH}{H} \!\!>\!\!\bigcirc\!\!<\!\! \underset{COOH}{H} & \underset{COOH}{H} \!\!>\!\!\bigcirc\!\!<\!\! \underset{H}{COOH} \\ H_2 \quad H_2 & H_2 \quad H_2 \end{array}$$

In the one case, the carboxyl groups are on the same side of the plane formed by the benzene ring, in the other case they are on opposite sides. The explanation is a transcription of that employed for maleïc and fumaric acids (cf. Chap. VII., p. 163); in fact, Baeyer recommends the terms *maleïnoid* and *fumaroid* for describing this kind of isomerism. As to benzene itself, we do not possess any definite information concerning the arrangement of its atoms in

[1] A long list of these is given in Meyer-Jacobson, *Lehrbuch*, II, 46 (1895). Two recent examples are : V. Meyer, *Ber. d. chem. Gesell.* **28**, 2776, 3195 (1895); and Hantzsch, *Ibid.* **29**, 958 (1896).

[2] *Ann. Chem.* (Liebig), **245**, 137, 158 (1888).

space, though numerous hypotheses have been advanced. Of these, but two call for mention here : that of Vaubel,[1] because its author is making a systematic attempt to develop it, and that of Sachse,[2] because, while geometrically excellent, it is incompatible with our views of valence, and may therefore lead to interesting developments of our valence conceptions. Unfortunately, the peculiarities of both of the stereochemical formulæ mentioned cannot be brought out without the aid of a model, so that no picture of them can be given here.

In conclusion, it may be worth while to risk a prophecy. As is well known, certain curious relationships obtain when we subject benzene derivatives to chemical action ; the nature of the substituents already present determines the position the entering substituent must take up. For example, the nitration of phenol leads to a mixture of the ortho- and para-nitro-phenols, whereas the nitration of benzoic acid leads to a meta derivative. This curious phenomenon thus far lacks an adequate explanation, though several attempts have been made to reduce it to law.[3] It is to be expected that the next progress in the elucidation of the benzene mystery will come from this field of investigation, — but what the results will be, mortal man cannot foresee.

Kekulé lived thirty years after immortalizing himself and his " theory " — thirty years devoted to incessant labor by himself and thousands of others. But in spite of the tremendous activity inspired by

---

[1] Cf. recent and current numbers of *J. prakt. Chem.*

[2] *Ztschr. physik. Chem.* 10, 228 (1892).

[3] Cf. Armstrong, *J. Chem. Soc.* 1887, 582.

the benzene theory, in spite of the heroic efforts of men like Ladenburg and Baeyer, the final answer has yet to be given to the tantalizing question : What is the constitution of C_6H_6? When will it come, and whence ?

CHAPTER IV

THE CONSTITUTION OF ACETOACETIC ETHER

THE chapter of organic chemistry which deals with the constitution of the substance called acetoacetic ether, prepared by Geuther in 1863, is second in its importance and extent only to that on benzene. At this day, when we are just emerging from the heated strife of diverging opinion, when the many new aspects of chemical reaction still startle us with their unexpectedness and variety, it is impossible fully to appreciate the far-reaching significance of the revolution that is being quietly effected. The mass of material collected is appalling in its vastness and complexity; but fortunately, during the past year or two, much insight has been gained into the nature of the puzzling problem, and a great part of the amassed facts are now only a matter of detail.

The early history of acetoacetic ether, though quite complicated, has but little interest to us now, for the complications arose from insufficient experimentation rather than from difficulties of interpretation. A full treatment of this portion of the subject has been given by Johannes Wislicenus,[1] a brief extract from which will suffice here.

Geuther's first paper remained practically unknown, for it was published in an almost inaccessible[2] quarter;

[1] *Ann. Chem.* (Liebig), **186**, 163–178 (1877).
[2] *Göttinger Anzeigen*, **1863**, p. 281.

it contained the announcement that when sodium is dissolved in pure ethyl acetate, the chief product is a sodium compound of the composition $C_6H_9NaO_3$. This sodium compound, treated with acids, gave an ethereal liquid $C_6H_{10}O_3$;[1] it reacted with alkyl iodides, replacing sodium by alkyl. Two years later, Frankland and Duppa[2] reported a series of products from the action of sodium on acetic ether after subsequent treatment with ethyl iodide. These products were ethylacetic acid ethyl ester, diethylacetic acid ethyl ester, and other compounds, which proved to be identical with the products obtained by Geuther by action of ethyl iodide upon his sodium compound $C_6H_9NaO_3$. The formation of these latter substances appeared simple enough, since Geuther had shown the steps of the reaction; but in order to explain the occurrence of the first two, Frankland and Duppa assumed that sodium forms several compounds with ethyl acetate, notably a mono- and a di-substitution product :

$$CH_3COOC_2H_5 + Na = CH_2NaCOOC_2H_5 + H$$
and
$$CH_2NaCOOC_2H_5 + Na = CHNa_2COOC_2H_5 + H$$

which then reacted quite normally with ethyl iodide :

$$CH_2NaCOOC_2H_5 + C_2H_5I = CH_2(C_2H_5)COOC_2H_5 + NaI$$
$$CHNa_2COOC_2H_5 + 2 C_2H_5I = CH(C_2H_5)_2COOC_2H_5 + 2 NaI$$

According to Frankland and Duppa, then, Geuther's sodium compound was only a part of the reaction product. As to its constitution, they found that

[1] Viz. acetoacetic ether.

[2] *Ann. Chem.* (Liebig), **187**, 217 (1865); **188**, 204, 328 (1866).

from its methyl and ethyl substitution products, methyl and ethyl acetone were obtained by saponification; and they accordingly expressed the opinion that the compound $C_6H_9NaO_3$ was really sodium acetone-carboxylic acid ethyl ester:

$$CH_3-CO-CHNa-CO-OC_2H_5$$

Geuther[1] did not accept any of these conclusions. He regarded the oil $C_6H_{10}O_3$ as an *acid* (he called it ethyldiacetic acid); he was unable to find any trace of Frankland and Duppa's supposed sodium-substituted acetic ethers; and finally he showed that the ethyl derivative of the sodium compound, which both he and the others had prepared, gave considerable quantities of butyric ester (ethylacetic ester) on simple heating with sodium ethylate. This last fact completely overthrew Frankland and Duppa's hypothesis; for sodium ethylate is always formed during the action of sodium upon acetic ether.

From this time on, until the classic researches of Wislicenus, the investigation was of an extremely desultory nature, consisting for the most part of theoretical disquisitions of a reprehensible character. Wislicenus clearly realized the necessity of working with pure materials, and of making sure of each step before passing on. His work extended over a period of several years, though the chief publications appeared in 1877.[2]

In order to properly understand Wislicenus's standpoint, the citation of his reasons for accepting the Frankland-Duppa formula for acetoacetic ether is de-

[1] *Ztschr. Chem.* **1868**, 58.
[2] *Ann. Chem.* (Liebig), **186**, 163 (1877); **190**, 257 (1877).

sirable. The passage in question, moreover, possesses considerable historic interest, as we shall see later.

"If acetoacetic ether is really ethyldiacetic acid, it is difficult to reconcile the fact that upon saponification by alkalies *it* yields carbonic acid and *acetone*, with the circumstance that the *esters* of this acid (made by action of alkyl iodides upon the sodium salt) do not also give acetone when saponified, but the corresponding alkylsubstituted acetones. Thus, Frankland and Duppa obtained from methyl acetoacetic ether, which, according to Geuther, can only be the methyl ester of ethyldiacetic acid, a methylated acetone (methyl-ethyl-ketone), from ethylacetoacetic ether, ethyl acetone (methyl-propyl-ketone), etc.

"The alkyl groups which take the place of the sodium atom in sodium acetoacetic ether (sodium ethyldiacetate according to Geuther) are found in the products of saponification *attached directly to carbon;* and in all probability the products first formed likewise contain them attached to carbon. But if this is the case, then the sodium atom of sodium acetoacetic ether (whose place of course is taken by the alkyl) is not attached as in salts; *i.e.* to oxygen, but directly to the carbon of the nucleus. The following empiric equations absolutely require analogous constitutions for acetoacetic ether and its alkyl substitution products:

1. $C_6H_{10}O_3 + 2\,NaOH = Na_2CO_3 + C_2H_5OH + CH_3COCH_3$
2. $C_6H_9(R)O_3 + 2\,NaOH = Na_2CO_3 + C_2H_5OH + CH_3COCH_2(R)$
3. $C_6H_8(R)_2O_3 + 2\,NaOH = Na_2CO_3 + C_2H_5OH + CH_3COCH(R)_2$

This analogy would be wanting if $C_6H_{10}O_3$ were ethyldiacetic acid, $C_6H_9(R)O_3$ its alkyl ester. Nor

is it apparent in what relation the di-alkyl acetoacetic ethers stand to these others if we accept
Geuther's view.   Perhaps this innate difficulty
explains why Geuther scarcely mentions the di-alkyl
acetoacetic ethers in his various papers, and avoids
their discussion altogether.

"As against Frankland's conception of $C_6H_9NaO_3$
as sodium-acetone-carboxylic-acid ethyl ester,

$$CH_3 - CO - CHNa - CO - OC_2H_5$$

we might have set objections to the assumed direct
substitution of hydrogen attached to carbon by
metals.   However, since we have become acquainted
with similar processes in other organic substances
(as $e.g.$ acetylene, nitroethane), such an assumption
loses its startling peculiarity.   At this day it no
longer surprises us that a carbon atom which is
directly united to two CO-groups can hold metals
in place of hydrogen.   For the union with one nitro
group suffices to produce a similar condition of
'electrochemical polarization' in the nitro-paraf-
fines — the simple loss of hydrogen in the acetylenes
achieves the same end."

The theoretical views just quoted, then, were Wislicenus's guide during his investigations.   His experimental results are well known, and require but a
short review.   First of all, he found that acetoacetic
ether under no circumstances reacted with more than
one atom of sodium — it contains but *one* directly
replaceable hydrogen atom, contrary to the views of
Frankland and Duppa.   Upon replacing the sodium
by an alkyl radical, Wislicenus found that the resulting alkyl-acetoacetic ether was again able to take

up an atom of sodium — it is therefore possible to *indirectly* replace *two* hydrogen atoms in acetoacetic ether by metals. Similar results were obtained from the action of sodium ethylate in place of sodium. If the sodium compounds are heated, decomposition takes place; sodium acetoacetic ether yields chiefly ethyl acetate; sodium ethyl acetoacetic ether yields ethyl butyrate; di-ethyl acetoacetic ether, which does *not* react with sodium ethylate in the cold, gives considerable quantities of di-ethyl acetic ether. These results completely cleared up the nature of Frankland and Duppa's reaction, as is readily seen.

Wislicenus was thus able to show that the "acetoacetic ether syntheses" are best carried out with pure materials, as presence of excess of sodium ethylate invariably complicates matters. The value of these syntheses was greatly enhanced by Wislicenus's study of the peculiar behavior of acetoacetic ether and its substitution products toward saponifying agents. Thus, dilute alkalies (aqueous potash or barium hydroxide) split these compounds largely as follows:

$$CH_3CO-C(X)(Y)-COOC_2H_5 + 2 H_2O = C_2H_5OH + H_2CO_3 \\ + CH_3CO-CH(X)(Y)$$

into alcohol, carbonic acid, and *ketones* (ketonic hydrolysis). Concentrated (preferably alcoholic) alkalies, on the other hand, break up the molecule at another point,

$$CH_3CO-C(X)(Y)-COOC_2H_5 + 2 H_2O = C_2H_5OH + CH_3COOH \\ + CH(X)(Y)COOH$$

into alcohol, acetic acid, and a *substituted acetic acid* (the so-called acid hydrolysis). By means of acetoacetic ether as a stepping-stone, then, we may prepare any desired ketone (containing one methyl group) and any desired acid of the acetic acid series by the proper choice of alkyl halides.

But these syntheses by no means limit the applicability of acetoacetic ether ; for this substance is like the hat from which the magician takes out a trunkful of oddities. If we act upon sodium acetoacetic ether with ethyl chloracetate, we can get aceto-succinic ether, and by hydrolysis succinic acid :

$$CH_3COCH \mid Na \mid COOC_2H_5 + CH_2 \mid Cl \mid COOC_2H_5 = NaCl +$$
$$CH_3CO - CHCOOC_2H_5$$
$$\mid$$
$$CH_2COOC_2H_5$$

$$\cdot CH_3CO \mid CHCOO \mid C_2H_5 + 3\,H_2O = 2\,C_2H_5OH + CH_3COOH$$
$$\mid CH_2COO \mid C_2H_5 \qquad\qquad + CH_2COOH$$
$$\mid$$
$$CH_2COOH$$

and by suitable combinations, a huge variety of di- and poly-basic acids. Another type of synthesis occurs when we act upon the sodium compounds with iodine; acetoacetic ether for instance, gives diacetosuccinic ether :

$$CH_3COCH \mid Na \mid COOC_2H_5 \qquad\quad CH_3CO - CHCOOC_2H_5$$
$$\qquad\qquad\qquad\qquad\qquad + I_2 = 2\,NaI + \qquad\qquad\mid$$
$$CH_3COCH \mid Na \mid COOC_2H_5 \qquad\quad CH_3CO - CHCOOC_2H_5$$

from which another endless series of compounds arises. Furthermore many acetoacetic ether derivatives, being ketones, may be reduced to the corresponding hydroxy-compounds :

$$CH_3COCH_2COOC_2H_5 + H_2 = CH_3CHOH - CH_2COOC_2H_5$$

Thus, acetoacetic ether yields β-hydroxybutyric ether. This on heating loses water, and gives crotonic ether :

$$CH_3CHOH - CH_2COOC_2H_5 = H_2O + CH_3CH = CHCOOC_2H_5$$

which enables us to pass to the unsaturated series of acids. In short, acetoacetic ether is our most valuable synthetic [1] reagent.

It would almost seem from these results of Wislicenus that the history of acetoacetic ether might have stopped then and there; but thousands of pages of controversy testify to the contrary. The most immediate and most important difference of opinion centred around the query: Are not the properties of acetoacetic ether better expressed by the isomeric formula of an oxy-crotonic ether :

$$CH_3 - C - (OH) = CH - CO_2C_2H_5 ?$$

The difference between the two formulæ is very slight ; one readily passes into the other by the simple shifting of a hydrogen atom. We may ask, then, what the arguments are in favor of the "oxy" formula. The question is complicated by many others, however, and though their solution was worked out simultaneously for the most part, a separate treatment of each is imperative. These questions are as follows :

1. Why is acetoacetic ether possessed of acid properties? Wislicenus speaks of "electrochemical polarization," of the influence of two carbonyl groups on the methylene between them :

$$CH_2 \begin{cases} CO - CH_3 \\ CO - OC_2H_5 \end{cases}$$

Is this a general law? Does any real significance attach to the reference to electrical phenomena?

2. Wislicenus asserts that when alkyls replace sodium, they do so literally ; they take the place of

---

[1] *E.g.* note Behrend's synthesis of uric acid (p. 106).

the sodium. Such a view has been almost axiomatic in chemistry ; but is it really and invariably true ? And if doubt be possible in this case, why must acetoacetic ether and its sodium compound necessarily have the same constitution ?

3. To what extent do the various condensation products of acetoacetic ether and its analogues correspond to one or the other formula ? Ketones condense differently from hydroxyl compounds, so that the possibility of a distinction seems evident.

4. The simple shifting of a hydrogen atom removes the difference between the " ketone " and the " oxy " formulæ. May we not assume such a shifting to take place during reactions, the direction of the oscillation to be determined by the reagents employed? Would not the two formulæ then become identical? Do we really know of any case where isomerism of the subtle nature demanded here actually exists?

Verily, a startling array of problems, well calculated to arouse differences of opinion. Their difficulty has resulted in one feature common to every solution proposed. The very fact that so many questions are asked about acetoacetic ether shows that they cannot easily be answered by a study of the compound itself. Therefore, each author has searched for analogies throughout the whole domain of organic chemistry. It is this feature that complicates the issue, and which has made it of such vital importance to the whole subject. It is this feature, moreover, which has caused most of the confusion and violence of the struggle ; for reasoning by analogy is a dangerous proceeding — all the more so when the mother-substance is Protean in its character.

I. THE ACTION OF SODIUM ON SIMILAR COMPOUNDS

To test the theory of "electrochemical polarization," Wislicenus searched for other compounds containing a methylene group adjacent to two carbonyls. Under his immediate inspiration, Conrad[1] found such a substance in the well-known malonic ether :

$$CH_2 \diagup^{CO-OC_2H_5}_{\diagdown CO-OC_2H_5}$$

which contains the desired grouping, and is second in importance only to acetoacetic ether itself. Malonic ether forms a sodium substitution product in much the same way as acetoacetic ether. A slight difference exists inasmuch as sodium malonic ether is decomposed by water, and sodium acetoacetic ether is not ; and further, that malonic ether can *directly* replace both its methylene hydrogen atoms by sodium, whereas acetoacetic ether cannot. Malonic ether is an important factor in the syntheses of fatty acids, since all the substituted malonic acids show a typical reaction of malonic acid itself ; viz., of splitting off carbon dioxide on heating, with formation of a member of the acetic acid series :

$$CH_2 \diagup^{COOH}_{\diagdown COOH} = CH_2 \diagup^{H}_{\diagdown COOH} + CO_2$$

Similar reactions toward sodium are shown by all substances possessing the group postulated by Wislicenus ;

$$-CO-\overset{|}{C}H-CO-$$

[1] *Ann. Chem.* (Liebig), **204**, 121 (1880).

Such are, for instance, *acetylacetone*:

$$CH_3CO-CH_2-CO-CH_3$$

acetone dicarboxylic ether:

$$CO \begin{cases} CH_2-CO-OC_2H_5 \\ CH_2-CO-OC_2H_5 \end{cases}$$

benzoylacetone:

$$C_6H_5CO-CH_2-CO-CH_3$$

benzoylacetic ether:

$$C_6H_5CO-CH_2-CO-OC_2H_5$$

besides many others which will be referred to later. A general law thus seems to hold here.

Yet Wislicenus stated only half of the truth; for in 1890 Freer[1] found that *acetone* itself,

$$CH_3-CO-CH_3$$

which certainly does not contain a methylene group situated between two carbonyls, is capable of forming a sodium derivative, whose reactions are similar in nearly every respect to those of sodium acetoacetic ether. This reaction of Freer's is a general one; it is shown by other ketones,[2] as well as by *acetic aldehyde*:[3]

$$CH_3-CO-H$$

The simple presence of a carbonyl group, then, suffices for the formation of sodium compounds, provided, of course, that a neighboring hydrogen atom be available.

— [1] *Amer. Chem. J.* 13, 308 (1890); 15, 582 (1893); 17, 1 (1895).
[2] *Ibid.* 19, 878 (1897).
[3] *Ibid.* 18, 552 (1896).

Wislicenus's theory of "polarization" thus resolves itself into a general property of ketones (and probably aldehydes); there is no connection with electrical phenomena as far as we know. On the other hand, it must be said that certain other so-called "negative" groups besides carbonyl possess the power of enabling neighboring hydrogen atoms to react with sodium. Thus, Victor Meyer[1] found that *benzyl cyanide:*

$$C_6H_5-CH_2-CN$$

gives a sodium compound which reacts with ethyl iodide, attaching the ethyl to the methylene group. Likewise, cyanacetic ether :[2]

$$CH_2 {\Large\langle} {{CN} \atop {CO-OC_2H_5}}$$

and malo-nitrile :[3]

$$CH_2 {\Large\langle} {{CN} \atop {CN}}$$

form similar sodium compounds. We thus have to deal with an undoubted influence of neighboring groups upon each other; but what the explanation may be, is at present unknown.

II. WHERE IS THE SODIUM IN SODIUM-ACETOACETIC ETHER?

The fact that such a question could ever have arisen marks one of the most important steps taken by chemistry. Frankland and Wislicenus found no

---

[1] *Ber. d. chem. Gesell.* **20**, 534, 2944 (1887).

[2] Henry, *Compt. rend.* **104**, 1618 (1887); Haller, *ibid.* 1626.

[3] Schmidtmann, *Ber. d. chem. Gesell.* **29**, 1171 (1896); Hesse, *Amer. Chem. J,* **18**, 723 (1896).

difficulty in answering it : the sodium, " whose place
is of course taken by the alkyl," is attached to carbon, since the alkyl groups are eventually found
there.   If the sodium were attached to oxygen, why
then the alkyl groups would be found there.   Upon
such reasoning as this the whole superstructure of
synthetic chemistry rests.

It is true, the validity of this argument had occasionally been questioned.   Thus, it is well known
that potassium cyanide gives *normal cyanides* with
alkyl iodides.:

$$R-C \equiv N$$

whereas silver cyanide gives *isocyanides :*

$$R-N=C=$$

Similar phenomena occur with nitrites and sulfites.
Yet this difficulty was avoided by the assumption of
different[1] constitutions of the potassium and silver
salts in question, or by a so-called rearrangement
of the product after its formation.

In 1887, however, Michael[2] found that sodium
acetoacetic ether, which, when treated with ethyl
iodide, gave a C-derivative,[3] gave a derivative of the
oxy-formula with chlorcarbonic ether :[4]

$$CH_3-C-(O-CO_2C_2H_5)$$
$$\|$$
$$CH-CO-OC_2H_5$$

---

[1] Cf. H. Goldschmidt, *Ber. d. chem. Gesell.* **23**, 253, 2180 (1890).

[2] And simultaneously Claisen (cf. *Ber. d. chem. Gesell.* **22**, 1763).

[3] Claisen (*Ann. Chem.* (Liebig), **277**, 162) has proposed that
when the entering radical becomes attached to carbon, it be called
a C-derivative, if to oxygen, an O-derivative.   This plan has been
universally adopted.

[4] *J. prakt. Chem.* [2], **37**, 473 (1887).

The literal interpretation of the formation of this *carbethoxy-acetoacetic ether* would lead to the oxy-formula for sodium acetoacetic ether; the equally literal explanation of the older syntheses of Frankland and Wislicenus would attach the sodium to carbon. If sodium acetoacetic ether is a single definite compound, as there is every reason to believe,[1] then one or the other of these syntheses is abnormal, and requires a special hypothesis.

One result stands out clearly from this experiment; viz., *that we cannot deduce the constitution of a substance from that of any of its derivatives, whose method of formation includes a metallic substitution product.* The most notable confirmation of this theorem is given by the history of the oximes (see Chap. VIII.). The question of the constitution of acetoacetic ether, therefore, is entirely independent of the structure of *sodium* acetoacetic ether, and all the arguments used to defend either the oxy or ketone formula of the free compound, which involve reactions of the sodium salt, are now ruled out of court. This independence cannot be shown more convincingly than by the fact that Claisen and Michael, two strong adherents of the ketone formula for the free ether, are also champions of the oxy structure of the sodium compound.

Two questions now arise: If acetoacetic ether and its sodium compound have different structures, how is one formed from the other? And if sodium acetoacetic ether can give both C- and O-derivatives, which of the two are we to consider abnormal?

---

[1] Cf. Elion, *Recueil trav. chim.* **3**, 240 (1884).

As to the first query, but little is known. However, it is largely a matter of indifference. The important question is, *What is* the constitution of the free compound and of its sodium salt? Our preconceived notions as to how a reaction *ought* to take place must wait until we know how it *does* take place. All such doctrines as affinity of sodium for oxygen, polarization, mutual attraction, and the like, valuable though they are to the individual as a spur to action and a guide to research, must not dominate the subject as a whole, to the exclusion of the real, and perhaps most simple, cause. In this connection it may be well to refer to the only theory of the formation of acetoacetic ether which rests upon an experimental basis, for it leads to the oxy-structure of the sodium salt. Claisen and Lowman[1] found that ethyl benzoate adds sodium ethylate :

$$C_6H_5C\underset{\diagdown}{\overset{\diagup OC_2H_5}{=}}O \quad + \ NaOC_2H_5 \ = \ C_6H_5C\underset{\diagdown ONa}{\overset{\diagup OC_2H_5}{\diagup OC_2H_5}}$$

This sodium addition product, brought into contact with ethyl acetate, reacts as follows :[2]

$$C_6H_5C\underset{\diagdown}{\overset{\diagup}{\begin{array}{c} OC_2H_5 \quad H \\ \hline OC_2H_5 \quad H \\ \hline ONa \qquad H \end{array}}} C - CO_2C_2H_5 = C_6H_5C(ONa) = CHCO_2C_2H_5 \\ + 2\,C_2H_5OH$$

giving alcohol and *sodium benzoylacetic ether.* The transfer of this fact to the action of sodium upon

---

[1] *Ber. d. chem. Gesell.* **20**, 651 (1887); **21**, 1154 (1888).

[2] Such condensations, produced by sodium ethylate, are extremely frequent and very important. Nearly all ketonic acids and poly-ketones are made by this means. The process has been extended chiefly by Claisen, W. Wislicenus, and Michael.

acetic ether is simple. Pure acetic ether is not attacked by sodium; the presence of a trace of alcohol is necessary; this forms sodium ethylate, which adds to a molecule of acetic ether; the addition-product then reacts with a second molecule to form sodium acetoacetic ether and alcohol; the alcohol produced reacts with more sodium, and thus the process goes on :[1]

$$CH_3C \begin{cases} OC_2H_5 \\ O \end{cases} + NaOC_2H_5 = CH_3C \begin{cases} OC_2H_5 \\ OC_2H_5 \\ ONa \end{cases}$$

$$CH_3C \begin{cases} \overline{OC_2H_5 \quad H} \\ OC_2H_5 + H \\ \overline{ONa \quad H} \end{cases} CCO-OC_2H_5 = 2 C_2H_5OH$$
$$+ CH_3C(ONa)=CHCO-OC_2H_5$$

The second question, which of the two forms of derivatives is abnormal, is more difficult to answer. So far but one solution has been proposed, which is far from being rigidly proved. Yet the ideas it contains have been so fruitful in directing research, and promise to so thoroughly modify our conceptions of the nature of chemical reactions, as to warrant a somewhat detailed consideration. This solution is the addition theory of Michael.[2]

The idea that many reactions proceed by addition and subsequent splitting off is not new. It was advanced by Baeyer[3] many years ago, and has been verified in many cases; we need only refer to the polymerization of aldehyde to aldol, with subsequent

---

[1] Claisen has given a summary of this theory, *Ann. Chem.* (Liebig), **297**, 92 (1897).

[2] *J. prakt. Chem.* [2], **37**, 486 (1888). *1888*

[3] *Ber. d. chem. Gesell.* **3**, 63 (1870).

loss of water to form crotonaldehyde, and to Perkin's reaction (see Chap. II.). The formation of esters from salts, on the other hand, had always been regarded as a process of direct interchange of the most pronounced type. Michael assumes as a starting-point that compounds containing metal and oxygen contain that metal attached to oxygen. (This assumption is not an integral part of the theory,[1] but the work of recent years seems to bring out this fact prominently.) Given a compound X-Y reacting with sodium acetoacetic ether, action may occur by replacement or by addition.[2] If by replacement, then the result is an O-derivative. If by addition, then addition occurs in such a manner that the group for which sodium has the greatest attraction (say Y) attaches itself to the carbon atom connected to NaO :

$$\begin{array}{c} Y \\ | \\ CH_3-C-ONa \\ | \\ C_2H_5CO_2-CH \\ | \\ X \end{array}$$

When, subsequently, NaY is split off, X remains attached to carbon :

$$\begin{array}{c} CH_3C{=}O \\ | \\ C_2H_5CO_2CHX \end{array}$$

---

[1] Nor is it universally accepted. It is still impossible to state *definitely* whether the sodium in the derivatives mentioned is attached to oxygen. The important fact to remember is that *no* matter *where* the sodium is situated, its position has nothing to do with the constitution of the free compounds.

[2] Or in both manners simultaneously. Cf. Claisen, *Ann. Chem.* (Liebig), **277**, 162 (1893); Wheeler and Boltwood, *Amer. Chem. J.* **18**, 381 (1896).

Whether addition or substitution is to take place depends upon circumstances at present unknown, but clearly capable of empirical determination. When addition occurs, the general course of reaction is as given above, since halogen is almost invariably one of the constituents of the reacting substance.

Michael has extended these considerations to all other cases where two forms of derivatives exist of a single mother-substance. Thus, potassium cyanide and silver cyanide may easily have the same structure; one may react with ethyl iodide by addition, the other by substitution. Michael himself has a " positive-negative " theory, from which he deduces the phenomena by the relative attraction of various groups for each other. But this theory, like so many others which are purely deductive, has failed in several cases[1] because of the insufficiency of its foundation ; and for the present it has no place in chemical history.

The addition theory, it is safe to predict, will play a very important part in the future development of chemistry. Yet a word of caution may not be amiss. If sodium acetoacetic ether reacts with ethyl iodide by addition, then the very similarly constituted sodium salts of the oxymethylene compounds (p. 83), *e.g.*

$$C_8H_{14} \begin{cases} C{=}CH{-}ONa \\ | \\ CO \end{cases}$$

ought, as far as we can see, to act in the same manner; yet sodium oxymethylene camphor gives an O-derivative[2] with ethyl iodide. We see, therefore,

---

[1] Cf. Marburg, *Ann. Chem.* (Liebig), **294**, 89 (1896).
[2] Claisen, *Ann. Chem.* (Liebig), **281**, 317 (1894).

that the theory still requires careful control by actual experiment. Then, too, the formation of diacetosuccinic ether from sodium acetoacetic ether (p. 66) is difficult to reconcile with the oxy-formula and the idea of addition.

But the chief obstacle to the growth of the addition theory lies deeper than these detailed exceptions. If we are going to explain reactions by means of addition products which we do not or cannot isolate, our explanation loses its definiteness. It becomes simply a *possible* explanation, and its conclusions are by no means binding. The recent history of Michael's hypothesis but serves to emphasize the necessity of critical caution. We owe most of our experimental knowledge of addition reactions to Nef, who has very lately based a far-reaching and fundamentally new conception of organic chemistry upon his results.[1] It is not time, nor is this the place, to attempt a criticism of this extension of the addition theory. But in the preliminary work leading up to his present standpoint, Nef has assumed as many as six[2] distinct intermediate products in certain reactions, not one of which supposititious stepping-stones has been isolated. It is clear that more definite ideas of reactions are necessary if we are to gain real insight into their mechanism. On the other hand, it must be admitted that transition periods are characterized by exaggeration, and we should not be hasty in our judgment of a valuable *working* hypothesis.

---

[1] *Ann. Chem.* (Liebig), **298**, 202 (1897).

[2] *Ibid.* **280**, 290 (1894); **287**, 347 (1895). Cf. V. Meyer, *Ber. d. chem. Gesell.* **27**, 3156 (1894).

III. THE CONDENSATION PRODUCTS OF ACETOACETIC ETHER

At a first glance it would seem as if the nature of these substances would immediately clear the matter up; for we have to decide between a ketone and a hydroxy compound, whose reactions are usually widely different. Upon second glance, however, it is apparent that the ordinary reagents for carbonyl will not help us, for on the basis of either formula we can arrive at identical products. Thus, acetoacetic ether adds hydrocyanic acid as well as sodium bisulfite ; but both of the addition products may be readily assumed to form from an oxy-crotonic ether ; *e.g.*

$$
\begin{array}{c}
CH_3 \\
\quad\diagdown C-OH \\
\quad\quad\| \\
\quad\quad CH \\
C_2H_5OCO
\end{array}
\; + HCN =
\begin{array}{c}
CH_3 \quad OH \\
\quad\diagdown C-CN \\
\quad\quad| \\
\quad\quad CH_2 \\
C_2H_5OCO
\end{array}
$$

In fact, it is due to the frequent formation of derivatives of the oxy-formula upon condensation that the controversy has continued so long. It is needless to give a list of examples here ; the discussion of a single case will suffice. Knorr[1] found that acetoacetic ether forms a hydrazone with phenylhydrazine, which, however, easily loses water and passes into phenyl methyl pyrazolone :[2]

$$
\begin{array}{c}
CH_3C\;|\;O \quad\quad H_2\;|\;N \\
\quad| \\
H_2C-CO + \\
\quad| \\
|\;OC_2H_5 \quad H\;|
\end{array}
\!\!
\begin{array}{c}
\diagdown \\
N-C_6H_5 \\
\diagup
\end{array}
=
\begin{array}{c}
CH_3C=N \\
\quad|\quad\quad\diagdown N-C_6H_5 \\
CH_2CO \diagup \\
\quad\quad + H_2O + C_2H_5OH
\end{array}
$$

[1] *Ann. Chem.* (Liebig), **238**, 147 (1886).

[2] This is the usual formula of this compound; two isomeric structures, however, are equally possible. Knorr, *Ann. Chem.* (Liebig), **279**, 186 (1893).

Nef[1] subjected the phenylhydrazone to a closer examination; he found that it can be oxidized to a compound containing two hydrogen atoms less, which is apparently an azo-body,[2] and concluded that acetoacetic ether phenylhydrazone is really phenyl-hydrazino-crotonic ether:

$$CH_3C-NH-NH-C_6H_5$$
$$\overset{\parallel}{C_2H_5OCOCH}$$

This formula is of course derived from the oxy-structure of acetoacetic ether, and is used by Nef as an argument for it.   However, Nef subsequently[3] showed in another connection that when phenylhydrazine reacts with *carbonyl* groups, it does so first by *addition*:

$$\begin{matrix} CH_3CO \\ | \\ CH_2CO-OC_2H_5 \end{matrix} + H_2N-NHC_6H_5 = \begin{matrix} O \, | \, H \\ CH_3C\!\!\!\diagdown| NH-NH-C_6H_5 \\ | \\ CH_2CO-OC_2H_5 \end{matrix}$$

This addition product, by loss of water, can readily pass into *either the hydrazone or the hydrazide* :

$$\begin{matrix} CH_3C\diagup \overset{HN-NH-C_6H_5}{\underset{OH}{\diagdown}} \\ | \\ C_2H_5OCOCH_2 \end{matrix} \begin{matrix} \nearrow \begin{matrix} CH_3C \\ | \\ C_2H_5OCOCH_2 \end{matrix}\diagdown N-NH-C_6H_5+H_2O \\ \\ \searrow \begin{matrix} CH_3C \\ \parallel \\ C_2H_5OCOCH \end{matrix}\diagdown HN-NH-C_6H_5+H_2O \end{matrix}$$

Thus no matter what the constitution of this condensation product, it does not help us to find the formula of the mother-substance.   Similar considerations apply to nearly all other condensations, so that they are of no avail.

---

[1] *Ann. Chem.* (Liebig), **266**, 70 (1891).

[2] See Freer, *Amer. Chem. J.* **21**, 23 (1899), for the real structure of this "azo-compound."

[3] *Ann. Chem.* (Liebig), **270**, 289, 333 (1892).

With one important exception, however. The condensation of acetoacetic ether with *ortho-formic ether*[1] can hardly be explained by any other than the *ketone* formula. The substances react to produce *diethoxy-butyric ether :*

$$CH_3C \underset{\underset{C_2H_5OCOCH_2}{|}}{[O]} + \begin{array}{c} C_2H_5O \\ C_2H_5O \\ C_2H_5O \end{array} {>} CH = CH_3C \underset{\underset{C_2H_5OCOCH_2}{|}}{<} \begin{array}{c} OC_2H_5 \\ OC_2H_5 \end{array} + HCO_2C_2H_5$$

It is difficult to see how such a product could be formed from an oxy-crotonic ether, except on the assumption of a variety of intangible addition reactions. If any one single reaction can be relied upon to settle the controversy, we have here a proof of the ketone formula of acetoacetic ether.

IV. MUST WE ASSUME A SHIFTING HYDROGEN ATOM IN ACETOACETIC ETHER ?

We have just seen that all derivatives formed from acetoacetic ether by addition or by means of its metallic compounds cannot help us decide upon its constitution, the only exception being the formation of diethoxy-butyric ether. To sum up our results in a general way, we should be inclined to call acetoacetic ether a ketone, chiefly from lack of valid reasons for considering it an oxy-compound. Attempts have not been wanting to ascertain the presence of a hydroxyl in the substance ; but the interpretation of the observed facts has varied with the standpoint of the observer. The following

---

[1] Claisen, *Ber. d. chem. Gesell.* **29**, 1005 (1896).

account must be largely a presentation of the personal opinions of the various investigators.

Geuther,[1] who was the first one to propose the oxy-crotonic formula, tried the action of acetyl chloride upon acetoacetic ether, as well as upon succinylosuccinic ether (p. 47). The latter readily forms a di-acetate; the former did not react until high temperatures were reached; hydrochloric acid was split off, but no acetate was formed. Yet in spite of this marked difference, Geuther[2] did not hesitate to declare the two substances to be similar in structure, viz. to contain hydroxyl groups. Considerably later, Nef[3] examined the action of acetic anhydride upon acetoacetic ether; he found that after prolonged boiling, about two per cent of the theoretical yield of O-acetyl acetoacetic ether was really obtained. Nef, too, interpreted this result to mean the presence of hydroxyl in the substance; many others, however, drew the opposite conclusion. Indeed, it hardly seems reasonable that a genuine hydroxy compound should give so small a quantity of ester after such energetic treatment.

The forced conclusions of Geuther and Nef are rendered still more improbable by the investigation of a series of substances containing the group postulated in oxy-crotonic ether, viz.,

$$-C-OH$$
$$\|$$
$$-CH$$

---

[1] Cf. Wedel, *Ann. Chem.* (Liebig), **219**, 87 (1883).

[2] *l.c.*, p. 119. Cf. *Ann. Chem.* (Liebig), **244**, 190 (1887).

[3] *Ann. Chem.* (Liebig), **276**, 212 (1893). Cf. v. Pechmann, *ibid.* **276**, 226 (1893).

Such substances are the so-called "formyl" or "oxy-methylene" compounds of Claisen and W. Wis-licenus, usually prepared by the sodium ethylate condensation of esters and ketones with formic ether; *e.g.* formyl- or oxymethylene-propionic ether :

$$CH_3 \qquad OC_2H_5 \qquad CH_3$$
$$HC - H + CHO \quad = \quad HC - CHO + C_2H_5OH$$
$$CO_2C_2H_5 \qquad\qquad CO_2C_2H_5$$

These formyl compounds (which closely resemble the *acetyl* compounds in their synthetic preparation and their general constitution) really have the iso-meric structure of oxymethylene derivatives : [1]

$$CH_3$$
$$C = CH - OH$$
$$CO - OC_2H_5$$

They react with benzoyl chloride, with acetic anhy-dride, with phosphorus *tri*chloride, to form the cor-responding derivatives with excellent yields; these are all undoubted hydroxyl reactions, and are given only slightly,[2] if at all, by acetoacetic ether. The conclusion thus seems justified that acetoacetic ether does not contain hydroxyl.

Here is where the matter rests to-day, as far as acetoacetic ether itself is concerned; an unbiassed summary of all the details given certainly points to the ketone formula for the free substance, and ren-ders the oxy-formula of its sodium salt probable. To a certain extent, both formulæ are thus justified and reconciled.

---

[1] *Ann. Chem.* (Liebig), **281**, 306 (1894).
[2] Cf. Nef, *ibid.* **276**, 238 (1893).

However, the participants in this history builded better than they knew; for the most important result of the whole agitation has arisen from the investigation of a proposed explanation which at the start had little in its favor. The hypothesis referred to is the well-known theory of pseudomerism, introduced by Baeyer,[1] developed by Laar,[2] Jacobson,[3] Hantzsch and Herrmann,[4] and others, and enriched by a variety of names, such as tautomerism, desmotropy, merotropy,[5] etc. This theory owes its origin to the fact that certain substances seem to possess more than one structure; *i.e.* they either give derivatives of two or more mother-substances, of which only one really exists; or their synthesis makes several formulæ equally probable for them. Examples of the first case are *isatine*,[6] which gives derivatives of the so-called "lactime" and "lactame" types:

Isatine
Lactime

Pseudo-isatine
Lactame

and *phloroglucine*,[7] which usually behaves like trioxy benzene, yet gives a tri-oxime derived from the tautomeric tri-keto-hexamethylene:[8]

[1] *Ber. d. chem. Gesell.* 16, 2189 (1883).   [2] *Ibid.* 18, 648 (1885).
[3] *Ibid.* 20, 1732 (1887).   [4] *Ibid.* 20, 2801 (1887).
[5] Michael, *J. prakt. Chem.* [2], 46, 208 (1892).
[6] Baeyer, *Ber. d. chem. Gesell.* 15, 2093 (1882).
[7] Baeyer, *ibid.* 19, 159 (1887).   [8] Cf. p. 54.

Examples of the second case are nitrosophenol, which can be prepared by the action of nitrous acid upon phenol,[1] as well as by action of hydroxylamine upon quinone : [2]

and phenyl tolyl formamidine,[3] which, whether prepared from formanilide and toluidine, or from formotoluide and aniline :

turns out to be one and the same substance.

This general phenomenon is now known indifferently by all the terms mentioned above, the most common designation being that of *tautomerism*. Much confusion of ideas has resulted hereby, so that the need[4] of a systematic designation has become pressing; but unfortunately, no uniform terminology has as yet been adopted. The name tautomerism was originally connected with a particular theory, viz.

[1] Baeyer and Caro, *ibid.* 7, 967 (1874).

[2] H. Goldschmidt, *ibid.* 17, 213 (1884).

[3] Cf. Wheeler, *Amer. Chem. J.* 19, 367 (1897); 20, 853 (1898); see also the diazo-amido compounds, p. 204.

[4] Cf. Claisen, *Ann. Chem.* (Liebig), 291, 46 (1896).

Laar's[1] hypothesis.   Laar assumed that the various formulæ of the substances mentioned above are but apparently different; they differ merely in the position of a single hydrogen atom, and become identical if we suppose the hydrogen atom in question to be continually oscillating between its two positions.[2] Laar's hypothesis, in this form, has not met with much favor; the more general view is that such substances have only *one* structure, but that during certain reactions they assume the other (pseudo-) forms.[3]

Now acetoacetic ether and oxy-crotonic ether may be regarded as a pair of tautomeric or pseudomeric substances.   We should merely have to decide which of the two formulæ appeared most frequently in reactions, and assign that formula to the free compound; the other would then be tautomeric.   This, indeed, was the commonly accepted explanation of the behavior of acetoacetic ether not so very long ago.   We have seen that the establishment of a separate individuality for sodium acetoacetic ether has done away with much of the necessity for any such hypothesis.

It must be admitted that the theory of pseudomerism does not satisfy one, for it is too intangible, too indefinite, too much of a glittering generality.   It is practically an evasion of the issue.   This seems to have been felt strongly by Nef, who has been a strenuous opponent of the idea.   His vigorous cham-

---

[1] *Ber. d. chem. Gesell.* **18**, 648 (1885); **19**, 730 (1886).

[2] Cf. Knorr, *Ann. Chem.* (Liebig), **279**, 186 (1894).

[3] Knorr, *l.c.* **293**, 97 (1896), has given a closer definition of tautomerism.   See also *l.c.* **303**, 133 (1898), footnote.

pionship of the oxy formula was probably due to a desire to explain all the reactions of acetoacetic ether without recourse to such a vague hypothesis. It cannot be said, however, at least not at present, that the addition theory is less vague[1] than tautomerism. Indeed, it would seem as if most of Nef's contentions in this field were erroneous. He assumed that *all*[2] 1-3 diketones are substances directly comparable; but we have seen that formyl compounds differ markedly from acetoacetic ether in their behavior, and thus cannot have a similar structure. As to tautomerism, Nef declared that the whole idea rested upon insufficient experimentation[3]; that no example of the subtle isomerism required by tautomerism is known.

But while the words were being penned, investigations of these very phenomena were going on in four different laboratories, with sufficiently startling results. The publications of Claisen and Wilhelm Wislicenus will stand as classics in the domain of pseudomerism, as models of experimentation in one of the most difficult fields of the science. Wislicenus[4] has found that formyl phenyl acetic ether exists in two modifications; the *a*-compound (a liquid) seems to have the formula:

$$\underset{\displaystyle CH(OH){=}\overset{\displaystyle C_6H_5}{\overset{\displaystyle |}{C}}{-}COOC_2H_5;}{}$$

[1] Cf. *Ber. d. chem. Gesell.* **25**, 1760 (1892); also *J. prakt. Chem.* [2], **46**, 189 (1892).

[2] *Ann. Chem.* (Liebig), **270**, 332 (1892); **276**, 240 (1893); **277**, 60, 74 (1893).

[3] *Ibid.* **266**, 137 (1891); **276**, 240 (1893); **287**, 353 (1895).

[4] *Ann. Chem.* (Liebig), **291**, 147 (1896).

it is a weak acid, and colors ferric chloride solution. The β-modification (solid) has the structure :

$$C_6H_5$$
$$CHO-H\overset{|}{C}-COOC_2H_5 \; ;$$

it is a much stronger acid than the *a*-compound, and does not color ferric chloride. These two substances can be readily converted into each other by simple heating or by use of solvents. High temperatures favor the *a*-form; solutions of either substance in chloroform soon contain chiefly *a*-, in alcohol, the β-. Claisen[1] has found, among other compounds,[2] that dibenzoyl acetyl methane and tribenzoyl methane each exist in similar modifications; the *a*-, or hydroxy, form :

$$C \diagdown \overset{\diagup C(OH)-CH_3}{\underset{\diagdown CO-C_6H_5}{-CO-C_6H_5}} \qquad C \diagdown \overset{\diagup C(OH)-C_6H_5}{\underset{\diagdown CO-C_6H_5}{-CO-C_6H_5}}$$

and the β-, or ketone, form :

$$H-C \diagdown \overset{\diagup CO-CH_3}{\underset{\diagdown CO-C_6H_5}{-CO-C_6H_5}} \qquad H-C \diagdown \overset{\diagup CO-C_6H_5}{\underset{\diagdown CO-C_6H_5}{CO-C_6H_5}}$$

In respect to behavior toward ferric chloride, Claisen's substances resemble Wislicenus's; the *a*- or *enol*[3] forms give a color reaction, the β- (*keto* or *aldo*) forms do not. Similarly, the *a*-modifications of all these substances are more stable at high temperatures. Yet

---

[1] *Ann. Chem.* (Liebig), **291**, 25 (1896).

[2] For a similar case, see Guthzeit, *Ann. Chem.* (Liebig), **285**, 35 (1895). Also Knorr, *ibid.* **293**, 70 (1896), where a number of cases are experimentally treated.

[3] The term *enol* is used as a class name for this kind of hydroxy-compound, *keto* and *aldo* for the corresponding isomeres.

a most curious difference exists between the Claisen and the Wislicenus bodies; for whereas Claisen's keto substances are *neutral,* and his enols *strong acids,* both of Wislicenus's are *acids,* the aldo being much stronger than the enol. How this contradiction of ideas will resolve itself, it is difficult to foretell; as matters stand to-day, the experimental evidence confirms both views. .

It has thus been experimentally demonstrated that substances exist which are capable of possessing *both* the enol and the keto structures :

$$- C(OH)=C - CO -$$
$$|$$

and

$$- CO - CH - CO -$$
$$|$$

Which of the two structures is the most stable, depends upon the nature of the appended radicals, upon the temperature, and upon the character of the solvent used. It is now merely a question of time until we shall be familiar with all the conditions necessary to determine one or the other constitution. Laar's hypothesis hereby loses any remnants of probability it possessed — *under given conditions we always have a definite substance before us.* At the same time, the theory of pseudo-forms receives an experimental basis; for at certain temperatures, or in certain solutions, when the enol is changing to keto, or *vice versa,* *both* substances are present; and if simple solution in an "indifferent" medium, such as alcohol or chloroform, is capable of effecting the transformation, compounds which can *react* with the substance must possess that ability to a still more marked degree.

We can readily see how the future of organic chemistry must wait upon the development of the theory which ascribes to the *reagent* the power of determining the constitution of the substance it attacks. It is not an extravagant comparison to place this new doctrine upon the same plane as Kekulé's benzene theory.

In conclusion, a reference to the " physical constants " of the substances which have been discussed may not be amiss. The physical constants have always been drawn into such extended controversies upon constitution. The general objection may be made to them that they represent an extrapolation, and that deductions from them are not legitimate when applied to substances whose *chemical* behavior is so uncertain (cf. p. 56). During the present controversy, however, the reliability of such physical determinations of constitution has been greatly strengthened ; for Brühl's [1] results from the value of the index of refraction, W. H. Perkin, Sr.'s [2] observations of magnetic rotation, and, finally, J. Traube's [3] new theory of molecular volumes, have severally and collectively agreed with the conclusions of purely chemical investigations.[4] Eventually, no doubt, when such close agreement between theory and observation will have become the rule, the problem of ascertaining

[1] *Ber. d. chem. Gesell.* 25, 366 (1892) ; *J. prakt. Chem.* [2], 50, 196 (1894) ; *Ann. Chem.* (Liebig), 291, 137, 217 (1896).

[2] *J. Chem. Soc.* 61, 836 (1892) ; *Ann. Chem.* (Liebig), 291, 185 (1896).

[3] Cf. *Ann. Chem.* (Liebig), 290, 43 (1895) ; *ibid.* 291, 188 (1896) ; *Ber. d. chem. Gesell.* 29, 1715 (1896).

[4] This applies also to the absorption of electric waves. Cf. Drude, *Ber. d. chem. Gesell.* 30, 957 (1897).

the structures of complicated substances will resolve itself into a glance into the spectrometer, controlled, it may be, by a specific gravity determination. Such an outcome is greatly to be desired ; but for the present the organic chemist must patiently bide by his furnaces and retorts, toiling in the sweat of his brow to wrest Nature's secrets from her.

CHAPTER V

THE URIC ACID GROUP

THE products of animal and vegetable metabolism have always exercised a peculiar fascination upon the chemical investigator. In the first place, these substances were the sole subject-matter of organic chemistry until 1828, when Wöhler proved the feasibility of organic synthesis. Secondly, many of these substances are of economic and medicinal importance, and the knowledge of their structure held out hopes of commercial gain. And finally, the processes whereby Nature evolves these complicated derivatives from the simple materials at her command are brought nearer their comprehension by a knowledge of the detailed organization of their molecules.

This last incentive has been peculiarly stimulating in the case of uric acid. This substance, a constant product of normal human physiology, assumes far greater significance in human pathology. Its excessive secretion is responsible for a large number of ailments, of which we need only mention rheumatism and stone in the bladder. It is no wonder, then, that uric acid has attracted considerable attention in the medical world, and that this interest has reacted upon the chemist. Incidentally, a number of compounds closely related to uric acid have been investigated because of their use as stimulants and drugs.

The first elaborate investigation of uric acid was published by Wöhler and Liebig[1] in 1838. This treatise is a classic in its experimental comprehensiveness. It brings a wealth of new material to light; but, owing to the state of chemical theory at the time, this material hardly serves us now in determining the structure of the acid. We shall therefore pass it by, for lack of space; but the reader who wishes fully to appreciate the greatness of these pioneers should not fail to peruse this masterpiece of their united genius.

The continuation of this investigation was taken up by Schlieper,[2] who added a number of new compounds to the already long list, and above all by Baeyer,[3] to whom we owe the first systematic relationship among the numerous derivatives. It is impossible to recount here the intricate experimental details, or even the marvellous chemical instinct, with which Baeyer unravelled his maze of facts. We must be satisfied with a synopsis of his main results.

Baeyer shows that uric acid and its host of derivatives may in their turn be regarded as derived from a compound $N_2C_4O_3H_4$, which he calls *barbituric* acid. Barbituric acid was found to break down into ammonia, carbon dioxide, and malonic acid when saponified;[4] Baeyer therefore looked upon it as malonyl urea:

$$\begin{array}{ccc} & CO - NH & \\ CH_2 & & CO \\ & CO - NH & \end{array}$$

[1] *Ann. Chem.* (Liebig), **26**, 241 (1838).
[2] *Ibid.* **55**, 251; **56**, 1 (1845).   [3] *Ibid.* **127**, 1, 199 (1863).
[4] *Ibid.* **130**, 143 (1864).

Barbituric acid yields a mono- and a di-brom substitution product, in which the bromine replaces hydrogen in the methylene group. If these bromine atoms are exchanged for hydroxyls, there are obtained *dialuric* acid:

$$\begin{array}{c} CO - NH \\ CH(OH) \quad \diagup \diagdown \quad CO \\ CH - NH \end{array}$$

(which may also be called hydroxybarbituric acid, or tartronyl urea), and dihydroxybarbituric acid:

$$\begin{array}{cc} CO - NH & CO - NH \\ C(OH)_2 \quad CO, \ or & CO \qquad CO + H_2O \\ CO - NH & CO - NH \end{array}$$

commonly known as *alloxan* (or mesoxalyl urea).

When dialuric acid is heated, a substance called hydurilic acid is formed (for an explanation of this, see below). Hydurilic acid, in its turn, when treated with nitric acid, breaks down into alloxan and either *violuric* or *dilituric* acids. Violuric acid, according to Baeyer, is nitroso-barbituric acid; it is now regarded as isonitroso-barbituric acid:[1]

$$\begin{array}{c} CO - NH \\ C=NOH \quad CO \\ CO - NH \end{array}$$

Dilituric acid is nitrobarbituric acid:

$$\begin{array}{c} CO - NH \\ HC(NO_2) \quad CO \\ CO - NH \end{array}$$

---

[1] It may also be looked upon as the oxime of alloxan. Ceresole, *Ber. d. chem. Gesell.* **16**, 1133 (1883).

The reasons for these formulæ are simple enough: reduction of both violuric and dilituric acids leads to an amido compound called *uramil;*[1] and uramil, when treated with nitrous acid, passes into dialuric acid. Dialuric acid being *hydroxy*barbituric acid, uramil must be *amido*barbituric acid:

$$
\begin{array}{c}
\text{CO-NH} \\
\text{HC-NH}_2 \quad\text{CO} \\
\text{CO-NH}
\end{array}
$$

and the formulæ of violuric and dilituric acids result as a combination of these facts with the empirical composition of the substances.

When uramil is treated with cyanic acid, the cyanate first formed rearranges itself as does ammonium cyanate, and a urea-derivative called *pseudo-uric* acid is the result. It receives this name because it has the empirical composition of uric acid, viz. $C_5N_4H_4O_3 + H_2O$; its constitution is:

$$
\begin{array}{c}
\text{NH-CO} \\
\text{CO} \quad \text{CH-NH-CO-NH}_2 \\
\text{NH-CO}
\end{array}
$$

Baeyer believed that if pseudo-uric acid could be made to lose one molecule of water, the resulting cyanamide[2] derivative would be uric acid, to which

---

[1] This had previously been prepared by Liebig and Wöhler, by the action of sulphur dioxide upon alloxan (thionuric acid is formed as an intermediate product).

[2] Urea, $NH_2-CO-NH_2$, passes by (indirect) loss of water into cyanamide, $N \equiv C-NH_2$, which in its turn takes up water (directly) to form urea.

he thus ascribed the following constitution :

$$CO \Big\langle \begin{array}{l} NH-CO \\ \quad\quad | \\ CH-NH-CO\equiv N \\ \quad\quad | \\ NH-CO \end{array}$$

It may be said in passing that Baeyer's efforts to effect
this synthesis of uric acid were in vain, and that
his formula was not generally adopted ; all the other
structures developed by him stand until this day.

To facilitate reference to these compounds, with
which the subsequent pages will deal frequently, the
following table given by Baeyer will be found very
convenient :

$N_2C_4O_3H_4$	Malonyl urea	Barbituric acid
$N_2C_4O_3H_3-OH$	Monohydroxy do. do.	Dialuric acid
$N_2C_4O_3H_2-(OH)_2$	Dihydroxy do. do.	Alloxan
$N_2C_4O_3H_2=NOH$	(Iso-)nitroso do. do.	Violuric acid
$N_2C_4O_3H_3-NO_2$	Nitro do. do.	Dilituric acid
$N_2C_4O_3H_3-NH_2$	Amido do. do.	Uramil
$N_2C_4O_3H_3-NH_2(CNOH)$	Amido-cyanate do. do.	Pseudo-uric acid

Baeyer has also given us an explanation of another
group of compounds, viz. alloxantine, hydurilic acid
(cf. above, p. 94), and violantine.   Alloxantine is
formed from a molecule each of alloxan and dialuric
acids by loss of two molecules of water :

$$N_2C_4O_3H_2(OH)_2 + N_2C_4O_3H_3(OH) = N_2C_4O_3H_2-O-N_2C_4O_3H_2 \\ + 2\,H_2O$$

Hydurilic acid is formed similarly from dialuric and
barbituric acids, by loss of only one molecule of
water :

$$N_2C_4O_3H_3(OH) + N_2C_4O_3H_4 = N_2C_4O_3H_3-N_2C_4O_3H_4 + H_2O$$

and violantine is a loose molecular combination of violuric and dilituric acids :

$$[N_2C_4O_3H_2(NOH) + N_2C_4O_3H_3(NO_2)]$$

None of these three compounds are of importance in connection with the constitution of uric acid.

Baeyer concluded his investigations with the synthesis of *hydantoïn*, one of the oxidation products of uric acid. Monobromacetyl urea is formed by action of bromine upon acetyl urea :

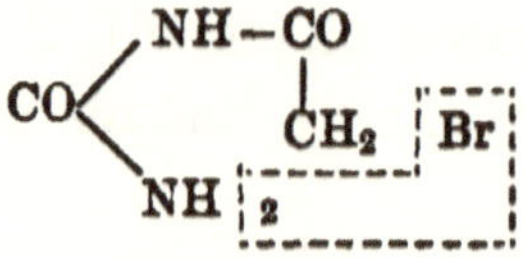

When heated, this loses hydrobromic acid as indicated, and hydantoïn is formed :

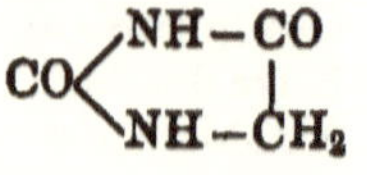

and is thus shown to be glycolyl urea. It may be regarded as a reduction product of *parabanic* acid or oxalyl urea :

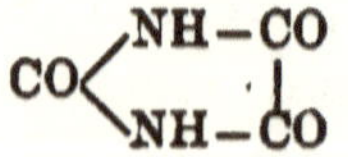

In the meantime, a number of other natural products were being brought into connection with uric acid. In particular, it was suspected that guanine and xanthine were related to uric acid, owing to the similarity in composition :

$C_5H_4N_4O_3,$ uric acid
$C_5H_5N_5O,$ guanine
$C_5H_4N_4O_2,$ xanthine

No facts were known which directly proved any con-
sanguinity between these three substances. In 1859,
Strecker[1] succeeded in converting guanine into
xanthine by means of nitrous acid :

$$C_5H_5N_5O + HNO_2 = C_5H_4N_4O_2 + N_2 + H_2O$$

The formulæ of the two compounds could therefore
be resolved :

$$C_5H_3N_4O(NH_2) \qquad C_5H_3N_4O(OH)$$
Guanine   Xanthine

Two years later,[2] he offered an explanation of the
relationship of xanthine, theobromine, and caffeïne.
Theobromine was to be regarded as dimethyl-, caffeïne as trimethylxanthine :

$$C_5H_2N_4O_2(CH_3)_2 \qquad C_5HN_4O_2(CH_3)_3$$
Theobromine   Caffeïne

Strecker partially proved this view by directly converting theobromine into caffeïne ; an attempt to
prepare the former from xanthine failed. Strecker
had not erred, however ; for many years later E.
Fischer[3] was able to effect the transition by a slight
modification of Strecker's method.

No notable advance in the history of uric acid
occurred until 1875, when Medicus suggested series
of structural formulæ for the whole group of substances. Curiously enough, though advanced upon
evidence hardly worth the name, these formulæ have
constituted the centre of discussion throughout the

---

[1] *Ann. Chem.* (Liebig), **108**, 141 (1859).
[2] *Ibid.* **118**, 170 (1861).   [3] *Ibid.* **215**, 253 (1882).

subsequent development of this complicated branch of structural chemistry.

Medicus[1] proposed the following constitution for uric acid :

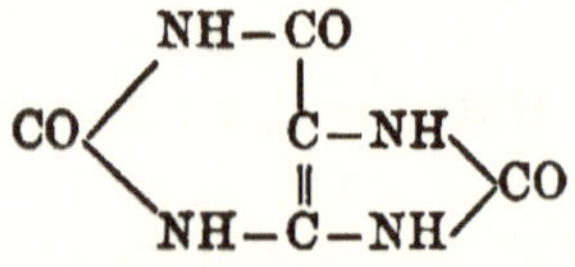

$$CO \Big\langle \begin{matrix} NH-CO \\ | \\ C-NH \\ \| \\ NH-C-NH \end{matrix} \Big\rangle CO$$

IIis reasons, briefly stated, were first that heating with acids gives glycocoll (amidoacetic acid), ammonia and carbonic acid; secondly, that as shown by Liebig and Wöhler, alloxan and urea are formed upon oxidation; and thirdly, that oxidation with potassium permanganate (in alkaline solution) gives allantoïn,[2] which readily passes into urea and glyoxylic acid.

Led on by the supposed close relation between uric acid and the substances just mentioned, Medicus ascribed similar structures to xanthine, guanine, etc.

Xanthine
$$CO \Big\langle \begin{matrix} NH-CO \\ | \\ C-NH \\ \| \\ NH-C-N \end{matrix} \Big\rangle CH \cdot$$

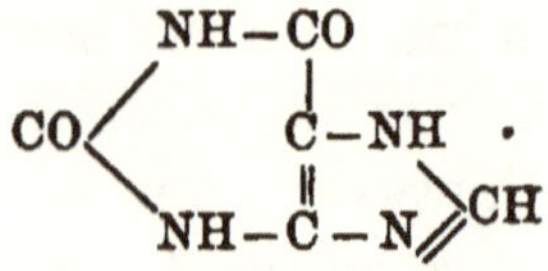

Theobromine
$$CO \Big\langle \begin{matrix} (CH_3)N-CO \\ | \\ C-NH \\ \| \\ (CH_3)N-C-N \end{matrix} \Big\rangle CH$$

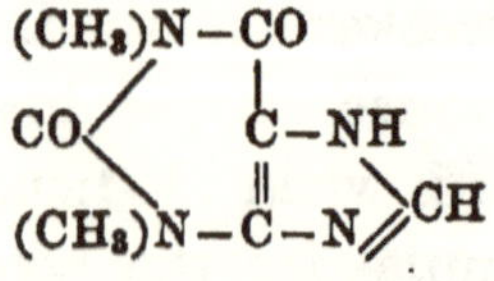

[1] *Ann. Chem.* (Liebig), **175**, 236 (1875).

[2] Allantoïn probably possesses the following structure:

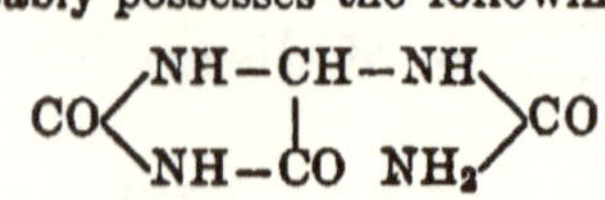

$$CO \Big\langle \begin{matrix} NH-CH-NH \\ | \\ NH-CO \quad NH_2 \end{matrix} \Big\rangle CO$$

viz. a di-ureïd of glyoxylic acid.

Caffeïne

$$
CO\Big\langle
\begin{array}{l}
(CH_3)N-CO \\
\quad\quad\quad\quad C-N(CH_3) \\
(CH_3)N-C-N
\end{array}
CH
$$

Guanine

$$
NH=C\Big\langle
\begin{array}{l}
NH-CO \\
\quad\quad\quad C-NH \\
NH-C-N
\end{array}
CH
$$

Hypoxanthine

$$
CH\Big\langle
\begin{array}{l}
NH-CO \\
\quad\quad\quad C-NH \\
N-C-N
\end{array}
CH
$$

For these formulæ absolutely no experimental evidence was adduced. This was of minor importance, however, since none of Medicus's proposals attracted attention. For uric acid, indeed, popular choice favored a formula advocated by Fittig in his well-known text-book:[1]

$$
CO\Big\langle
\begin{array}{l}
NH-C-NH \\
CO\langle \quad | \quad \rangle CO \\
NH-C-NH
\end{array}
$$

Again the prosecution of the problem came to a standstill for a considerable period. In 1882, however, an elaborate experimental investigation of *caffeïne* was published; the name of its author — Emil Fischer — guaranteed its contents.[2] Fischer's reason for attacking this substance first seems to have been the hope of finding a pathway to it from uric acid. At any rate, it was not hard (for Fischer) to unravel the structure of caffeïne, and subsequently to apply his new experiences to uric acid itself.

[1] *Grundriss der Organ. Chem.* 10th edition, p. 309.
[2] *Ann. Chem.* (Liebig), 215, 253 (1882).

The following facts give the key to Fischer's caffeine formula :

$$
\begin{array}{l}
(CH_3)N - CH \\
\qquad\quad \| \\
CO \diagdown \quad C - N(CH_3) \\
\qquad\qquad | \qquad \diagup CO \\
(CH_3)N - C = N
\end{array}
$$

a formula differing but slightly from that given by Medicus. In the first place, treatment of caffeine with chlorine in water solution yields *dimethyl*-alloxan and *monomethyl*-urea (a process of simultaneous oxidation and hydrolysis). Therefore nine of the ten hydrogen atoms in caffeine occur in the form of *three methyl* groups.

Secondly, the *tenth* hydrogen atom may be directly replaced by chlorine or bromine (forming chlor- or brom-caffeine). On heating bromcaffeine with ammonia we can get *amino*-caffeine, with caustic potash *hydroxy*-caffeine.

In the third place, hydroxycaffeine contains a double bond somewhere among its carbon atoms. We know this because it takes up a molecule of bromine, and because further the bromine atoms thus added may be exchanged for ethoxyl groups (OC_2H_5), by the simple action of alcohol. The resulting product is called *di-ethoxy-hydroxycaffeïne*.

Fourthly, diethoxyhydroxycaffeine is easily converted into *caffuric* acid. The first step consists in eliminating one molecule of methylamine and two of alcohol by boiling with concentrated hydrochloric acid, whereby a substance called *apocaffeïne* is formed. Apocaffeïne loses carbon dioxide when boiled with water, and passes into caffuric acid. Caffuric acid may also be obtained by moderately oxidizing caf-

feïne itself with chlorine.   Caffuric acid, on its part,
may be hydrolyzed[1] into mesoxalic acid, methylamine,
and monomethylurea.

Reducing agents, *e.g.* hydriodic acid, abstract an
oxygen atom from caffuric acid, forming *hydrocaf-*
furic acid, which is thus the corresponding derivative
of tartronic acid.   Hydrocaffuric acid is decomposed
on boiling with barium hydroxide, yielding methylamine, carbon dioxide, and *methylhydantoïn.*   The
monomethyl-urea residue of caffeïne is therefore
united to the rest of the molecule in the following
manner :

$$-\underset{|}{\overset{}{C}}-N(CH_3) \atop -\underset{}{\overset{}{C}}-N} >CO$$

this being the skeleton of methylhydantoïn (cf.
p. 97).   This is one of the most important data in
fixing upon the final structure of caffeïne.

A sixth link in the chain starts from diethoxy-
hydroxycaffeïne. When this substance is decomposed
with hydrochloric acid, the substance *hypo*caffeïne
accompanies the main product apocaffeïne.   Hypo-
caffeïne differs from the latter in its composition by
a molecule of carbon dioxide.   It loses carbon dioxide on boiling with bases, and is converted in *caffo-*
line.   One of the oxidation products of this substance
is dimethyloxamide, which contains its methylamine
groups attached to *two adjacent* carbon atoms.   Two
formulæ are thereby made possible for caffoline :

I.

$$\begin{array}{c} N(CH_3) \\ CH \diamondsuit CO \\ (CH_3)NH-CO \quad NH \\ (1) \end{array}$$

II. $$\begin{array}{c} HO.HC-N(CH_3) \\ | \quad >CO \\ (CH_3)NH-C=N \\ (1) \end{array}$$

---

[1] With basic lead acetate.

These formulæ differ in the manner in which nitrogen atom (1) is attached to the rest of the molecule. But I. is excluded because of the close relation of caffoline to hydrocaffuric acid, which latter we have just seen to contain the grouping II.

Fischer combines all of these rather complicated facts into the caffeïne formula already given. It will suffice for our present purposes to append the corresponding structure of hydroxycaffeïne :

$$
\begin{array}{l}
(CH_3)N-C-OH \\
\qquad\qquad \| \\
CO\big< \qquad C-N(CH)_3 \\
\qquad\qquad | \\
(CH_3)N-C{=}N\big>CO
\end{array}
$$

Since we know caffeïne to be *tri*-methylxanthine (p. 98), it is not difficult to formulate the mother-substance xanthine :

$$
\begin{array}{l}
(a)\ NH-CH \\
\qquad\qquad \| \\
CO\big< \qquad C-NH\ (c) \\
\qquad\qquad | \\
(b)\ NH-C{=}N\big>CO
\end{array}
$$

Theobromine is dimethylxanthine; as three places are open to occupation by but two alkyl radicals, it becomes necessary to determine their relative position. One methyl group is located in (c), for oxidation of theobromine gives methylurea. The other methyl occurs at (a); because if we replace the remaining imide hydrogen by an *ethyl* radical, we obtain *homologous* apo- and hypo-caffeïnes (ethyl-theobromine derivatives). In all of the above metamorphoses of the caffeïne molecule, the methyl group at (b) retains its place; as the ethyl group does likewise in ethyl-theobromine, the second methyl in theobromine must be attached to nitrogen atom (*a*).

The investigation of uric acid — in particular the establishment of a genetic connection between this substance and the xanthine series — next engaged Fischer's attention. The action of phosphorus pentachloride upon monomethyl-uric acid furnishes two new substances :

$$CH_3 . C_5N_4(OH) . Cl_2$$

$$CH_3 . C_5N_4 . Cl_3$$

formed by replacement of two, respectively three hydroxyl groups in the mother-substance by chlorine. These halogen derivatives manifest a number of reactions in which the chlorine atoms are replaceable. The resulting products, which need not occupy our attention here, may all be regarded as substitution-products of a compound $(CH_3)C_5N_4H_3$, which Fischer calls *methylpurin*. Henceforth, all the various substances derived from or related to uric acid may be considered *purin* derivatives :

$C_5N_4H_4$,   purin
$C_5N_4H_4O_3$,   trioxy-purin, uric acid
$CH_3 . C_5N_4H_3O_3$,   trioxymethyl-purin, methyluric acid
$(CH_3)_2C_5N_4H_2O_3$,   trioxydimethyl-purin, dimethyluric acid

The chlorine compounds above are known as dichloroxy-methyl-purin and trichlor-methyl-purin respectively. The importance of the purin series will appear subsequently.

The investigation of the various methyl-uric acids soon demonstrated the untenability of Fittig's formula (p. 100). The existence of a tetramethyl-uric acid, indeed, proved the presence of four imide groups in the acid ; but the two urea residues are not symmetrically arranged, and so Fittig's structure falls to the ground. In the first place, two[1] isomeric *mono*-methyluric acids exist; one of these may be oxidized to alloxan and methyl-urea, the other to methyl-alloxan and urea. This shows that the former is methylated *outside* the alloxan ring, an impossibility under Fittig's conception. In the second place, the four methyl groups in tetramethyluric acid are attached to nitrogen, because hydrolysis of this substance yields methylamine and *no* ammonia. And thirdly, the oxidation of dimethyluric acid produces cholestrophan (dimethyl-parabanic acid), which in its turn is an oxidation product of dimethylalloxan. Since the two methyl groups in this dimethyluric acid thus turn up in the alloxan ring, the remaining imide groups are outside of this ring. These various facts lead conclusively to the formula for uric acid originally proposed by Medicus:

$$\begin{array}{c} NH-CO \\ CO< \quad | \quad C-NH \\ \quad \quad \| \quad \quad >CO \\ NH-C-NH \end{array}$$

Uric acid may thus be looked upon as a di-ureïd of (condensation product of *two* molecules of urea with) trioxy-acrylic acid:

$$\begin{array}{c} HO-CO \\ | \\ C-OH \\ \| \\ HO-C-OH \end{array}$$

This acid is as yet unknown; but the knowledge that it forms the skeleton of uric acid made the

[1] *Four* are theoretically conceivable. They are all known. Cf. v. Loeben, *Ann. Chem.* (Liebig), **298**, 181 (1897).

synthesis of that substance a concrete possibility. And so the attempts to artificially prepare this complicated compound form the logical as well as the historical sequence of Fischer's analytic triumphs.

In part, the synthesis of uric acid antedates the investigations we have just scrutinized.   Horbaczewski,[1] in 1882, had succeeded in obtaining very small quantities of this substance by heating urea with glycocoll (amido-acetic acid).   The mechanism of the reaction is by no means clear, and the amount of uric acid it furnishes decidedly unsatisfactory. But by taking into account the relation of the compound to dioxyacrylic acid, and heating the very similar substance trichlor-lactic acid : [2]

$$
\begin{array}{c}
\text{HO} - \text{CO} \\
| \\
\text{H} - \text{C} - \text{OH} \\
| \\
\text{Cl} - \text{C} - \text{Cl} \\
| \\
\text{Cl}
\end{array}
$$

with urea, the yield of the desired substance was considerably increased.

We owe a much more complicated, but much more important, synthesis of uric acid to Behrend.[3]   When urea and acetoacetic ester are condensed, the first product, *uramidocrotonic ester :*

$$
\text{CO}
\begin{cases}
\text{NH} - \text{C} - \text{CH}_3 \\
\quad\quad \| \\
\quad\quad \text{CH} \\
\text{NH}_2 \quad \text{CO}_2 . \text{C}_2\text{H}_5
\end{cases}
$$

[1] *Monatsh.* **1882**, 796 ; **1885**, 356.
[2] *Ibid.* **1887**, 201, 584.
[3] *Ann. Chem.* (Liebig), **229**, 1 (1885).

loses alcohol upon saponification and forms a substance called *methyl-uracil* :

$$\text{CO} \underset{NH-CO}{\overset{NH-C-CH_3}{<}} \overset{\|}{CH}$$

It is the methyl derivative of the hypothetical compound uracil :

$$\text{CO} \underset{NH-CO}{\overset{NH-CH}{<}} \overset{\|}{CH}$$

which Behrend assumes to be the mother-substance of this series. Strong nitric acid attacks methyl-uracil; a nitro group is introduced, and the methyl group becomes oxidized to carboxyl :

$$\text{CO} \underset{NH-CO}{\overset{NH-C-COOH}{<}} C-NO_2$$

forming nitro-uracil carboxylic acid. This loses carbon dioxide when boiled with water, forming nitro-uracil :

$$\text{CO} \underset{NH-CO}{\overset{NH-CH}{<}} C-NO_2$$

Reducing agents convert this substance into the corresponding amido-uracil, but as a rule the reaction does not stop there; the amido group is replaced by hydroxyl, and oxy-uracil results :

$$\text{CO} \underset{NH-CO}{\overset{NH-CH}{<}} C-OH$$

Another formula is possible for this substance, tautomeric with the one given, but the reactions of oxyuracil render it less likely than the former. Oxyuracil is isomeric with barbituric acid, and may thus be called *isobarbituric* acid.

Isobarbituric acid proved rather an impassable barrier; and it was not until four years[1] later that Behrend was able to surmount the innumerable obstacles which blocked the way to his goal. Oxidation of oxyuracil (bromine water was the reagent which made this operation feasible) leads to *di*oxyuracil or *isodialuric* acid:

$$CO\begin{cases} NH-C-OH \\ \quad\quad \| \\ \quad\quad C-OH \\ \quad\quad | \\ NH-CO \end{cases}$$

And this, finally, condenses with urea to yield the long-sought uric acid:

$$CO\begin{cases} NH-C-\overline{OH \quad H}|NH \\ \quad\quad \| \quad\quad + \quad >CO \\ \quad\quad C-\underline{OH \quad H}|NH \\ \quad\quad | \\ NH-CO \end{cases} = CO\begin{cases} NH-C-NH \\ \quad\quad \| \quad\quad >CO \\ \quad\quad C-NH \\ \quad\quad | \\ NH-CO \end{cases} + 2\,H_2O$$

An elaborate comparison of synthetic uric acid with the natural substance proved the complete identity of the two.

This is where the history of uric acid stands to-day, or rather, stood yesterday. The various text-books of organic chemistry more or less completely outline our subject, as we have just seen it. But science makes rapid strides, and here, as elsewhere, has out-

[1] *Ann. Chem.* (Liebig), **251**, 235 (1888).

stripped the photographer with his cumbersome text-book camera. And so it has come to pass that the man who has taught us so much is now teaching us more. Fischer's syntheses in the uric acid group have led him to revise a considerable portion of his previous doctrines.

His first work in this new direction was a synthesis of uric acid itself. This is hardly the place for personal comments, but it seems characteristic of many of Fischer's achievements that he succeeds where others have tried and failed; and the admiration which this success commands is but heightened by the names of his predecessors. Liebig and Wöhler had correctly surmised the relation of uramil to uric acid; but their efforts to unite uramil and cyanic acid proved abortive. Baeyer, by a slight modification of their method, effected this union; but he was unable to dehydrate the resulting pseudo-uric acid. Fischer found the missing reagent; pseudo-uric acid gives up a molecule of water to molten oxalic acid,[1] and the dreams of Liebig, Wöhler, and Baeyer are realized. It should be pointed out that this new synthesis is simpler than the older ones, and that above all it is available for the syntheses of numerous homologous derivatives. Malonic acid condenses with urea and forms barbituric acid. Barbituric acid is converted into violuric acid by nitrous acid. Reduction of violuric acid leads to uramil. Uramil and potassium cyanate yield pseudo-uric acid. And, finally, pseudo-uric acid loses water and forms uric acid itself. It will be readily seen that by em-

---

[1] *Ber. d. chem. Gesell.* **28**, 2473 (1895).

ploying substituted ureas in the above series, substituted uric acids are obtainable.

Once the conversion of pseudo-uric acid was proved
possible it did not take Fischer long to find an easier
method; the dehydration occurs on simple boiling
with dilute mineral acids.[1] The application of this
simple process to some methyl-pseudo-uric acids has
led to startling results.   But first a word about the
nomenclature of these substances.   Wherever synthetic chemistry obtains a foothold, ordinary methods
of naming compounds soon prove inadequate and
confusing.   Fischer proposes[2] to base all names upon
the fundamental *purin* nucleus which he employed
years ago (cf. p. 104).   The nine atoms of this
nucleus are to be numbered thus:

$$
\begin{array}{c}
1\,\mathrm{N}\text{--}\mathrm{C}\,6 \\
2\,\mathrm{C}\quad 5\,\mathrm{C}\text{--}\mathrm{N}\,7 \\
\qquad\qquad\qquad \mathrm{C}\,8 \\
3\,\mathrm{N}\text{--}\mathrm{C}\text{--}\mathrm{N}\,9 \\
4
\end{array}
$$

and the place of the substituents indicated by placing
the proper number before them.

When 1, 3, 7-trimethylpseudo-uric acid:

$$
\begin{array}{l}
\mathrm{CH_3-N-CO}\quad \mathrm{CH_3} \\
\mathrm{CO}\!\!< \quad \mathrm{CH-N-CO-NH_2} \\
\mathrm{CH_3-N-CO}
\end{array}
$$

is dehydrated with dilute hydrochloric acid, the
resulting trimethyluric acid *proves to be identical
with hydroxycaffeïne.*   It will be remembered that

[1] *Ber. d. chem. Gesell.* **30,** 559 (1897).   [2] *Ibid.* p. 557.

Fischer had come to the conclusion that xanthine and uric acid contained *different* carbon chains. Here was a surprisingly close relation between caffeïne, a simple xanthine homologue, and trimethyluric acid, standing in a similar position to uric acid. If any doubts remained in Fischer's mind, they were speedily relieved by the discovery that direct alkylation converts hydroxycaffeïne into tetramethyluric acid. And, finally, it is possible directly to prepare hydroxycaffeïne by methylating uric acid.

Hydroxycaffeïne, then, is trimethyluric acid, and the former conclusions must be revised upon the basis of its new formula:

$$
\begin{array}{l}
CH_3-N-CO \\
\quad\diagup \quad | \\
CO \quad\quad C-N-CH_3 \\
\quad\diagdown \quad \| \quad >CO \\
CH_3-N-C-NH
\end{array}
$$

Caffeïne, xanthine, and guanine can no longer be construed as before; the best expression of their behavior is given by the structures proposed by Medicus, but heretofore discredited upon Fischer's authority:

Caffeïne
$$
\begin{array}{l}
CH_3.N-CO \\
\quad\diagup \quad | \\
CO \quad\quad C-N-CH_3 \\
\quad\diagdown \quad \| \\
CH_3-N-C-N\!\!>\!CH
\end{array}
$$

Xanthine
$$
\begin{array}{l}
NH-CO \\
\quad\diagup \quad | \\
CO \quad\quad C-NH \\
\quad\diagdown \quad \| \\
NH-C-N\!\!>\!CH
\end{array}
$$

Guanine
$$
\begin{array}{l}
NH-CO \\
\quad\diagup \quad | \\
NH=C \quad\quad C-NH \\
\quad\diagdown \quad \| \\
NH-C-N\!\!>\!CH
\end{array}
$$

The formula of theobromine is rendered uncertain; to be sure, one methyl group is attached to the hydantoïn complex, but the position of the other in the alloxan ring must be determined afresh (cf. below).

The old question of effecting a transition from uric acid to xanthine thus finds a simple solution. Historically, however, it had been accomplished two years previously by Fischer,[1] during the course of the synthesis of caffeïne. 1, 3-dimethyluric acid passes into chlor-theophyllin when treated with phosphorus pentachloride, and reduction of the latter leads to *theophyllin*, which is 1, 3-dimethylxanthine:

$$
\begin{array}{ccc}
\text{CH}_3\text{-N-CO} & \text{CH}_3\text{-N-CO} & \text{CH}_3\text{-N-CO} \\
\diagup \quad | & \diagup \quad | & \diagup \quad | \\
\text{CO} \diagdown \quad \text{C-NH} \diagdown & \text{CO} \diagdown \quad \text{C-NH} \diagdown & \text{CO} \diagdown \quad \text{C-NH} \diagdown \\
\quad \| \quad \diagup\text{CO} & \quad \| \quad \diagup\text{C-Cl} & \quad \| \quad \diagup\text{CH} \\
\text{CH}_3\text{-N-C-NH} & \text{CH}_3\text{-N-C-N} & \text{CH}_3\text{-N-C-N} \\
\text{1, 3-dimethyluric acid.} & \text{Chlortheophyllin.} & \text{Theophyllin.}
\end{array}
$$

Methylation of theophyllin[2] leads to caffeïne, and thus a complete synthesis of the latter is achieved.

But to recount the work of the last two years would need a volume by itself. It must suffice here to briefly mention a few of the more important facts — indeed, as yet we have not enough perspective to measure their full value. Complete syntheses are at hand for xanthine,[3] guanine, and a number of related substances; adenin,[4] a decomposition product of nucleines, hypoxanthine, a similar substance occurring in extract of meat, and hetero- and para-

[1] *Ber. d. Chem. Gesell.* **28**, 3135 (1895).

[2] Kossel, *Ztschr. physiol. Chem.* **13**, 305 (1889).

[3] *Ber. d. Chem. Gesell.* **30**, 2226 (1897).

[4] *Ibid.* **31**, 104 (1898).

xanthine,[1] isomeric mono-methyl xanthines of some physiological importance. Theobromine, too, has been prepared synthetically from 3, 7-dimethyl-pseudo-uric acid,[2] and is therefore 3, 7-dimethyl-xanthine:

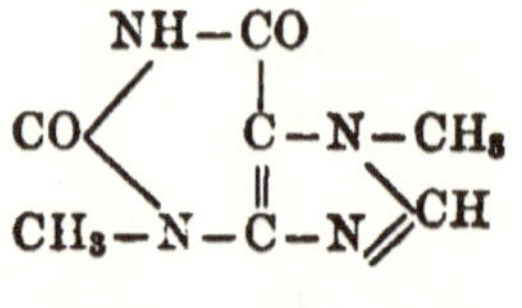

$$
\begin{array}{c}
NH-CO \\
CO\Big\langle \quad \begin{array}{c} C-N-CH_3 \\ \| \\ CH_3-N-C-N \end{array} \Big\rangle CH
\end{array}
$$

instead of 1, 7-dimethylxanthine as heretofore.

The satisfactory resolution of the uric acid group closes one of the most noteworthy chapters in the history of chemistry. It is difficult to estimate the full significance of this work, or to set a limit to its value for the physiology of the future. It is clear that an adequate comprehension of cellular processes in the human body must wait upon thorough chemical knowledge of the products of cellular activity. The inspiring nature of this chemical problem cannot be better exemplified than by appending the names of those who have given their best thoughts to it — Liebig, Wöhler, Baeyer, Emil Fischer.

[1] *Ber. d. Chem. Gesell.* **30**, 2400 (1897).
[2] *Ibid.* **30**, 1839 (1897).

I

CHAPTER VI

THE great family of the sugars, starches, cellu-
loses, and gums, familiarly known as the carbohydrates, constitutes one of the earliest recognized
and most important groups of organic substances.
They play a decisive part in the animal and vegetable organism, forming a large portion of its food
and its tissues. More capital and labor are involved
in their production and application than in any other
branch of chemical technology ; it is only necessary
to refer to the manufacture of cane and beet sugar,
of starch, of fermented liquors, of paper and other
cellulose preparations, and to the distillation of
wood. The amount of work devoted to the chemical investigation of the carbohydrates is commen-
surate with this importance. The literature of
organic chemistry contains innumerable treatises on
the behavior and constitution of individual members of the series. But the actual contribution
toward a systematic knowledge of composition and
constitution, of genetic relations among and synthetic preparation of these compounds, during the
first eighty-odd years of the century, is very small
indeed. It is due to the efforts of two men, H.
Kiliani and Emil Fischer, that the former chaos of
isolated facts has become one of the most beautifully
organized chapters of science. Step by step, the

114

structure of these substances has been unravelled; and step by step, they have been built up from the simplest of materials to the full complexity of grape-sugar and its congeners. Not only have we learned the inner anatomy of Nature's products, but numerous new carbohydrates have been added to the list until to-day their number is legion. If less than ten years have witnessed the complete subjugation of the sugars to order and system, what may we not hopefully look for in Nature's box of mysteries?

The first attempt to give a precise constitution to a particular sugar was made by Baeyer.[1] This sugar was the so-called *methylenitane*, discovered by Butlerow.[2] The fact that the composition of the ordinary sugars was expressed by the formula: $C_6H_{12}O_6$, or $(C-H_2O)_6$, naturally suggested that these compounds might be obtained by simple polymerization of any substance possessing the composition $(CH_2O)_x$; such, *e.g.*, are formaldehyde, and its polymer trioxymethylene $(CH_2O)_3$. Butlerow found that by adding lime water to trioxymethylene a sweet sugarlike substance was formed, to which he gave the name methylenitane. Baeyer was of the opinion that its formation could be explained by simple condensation of six molecules of formaldehyde, thus:

$$\overset{1}{CH_2O} \mid \overset{2}{H-CHO} \mid \overset{3}{H-CHO} \mid \overset{4}{H-CHO} \mid \overset{5}{H-CHO} \mid \overset{6}{H-CHO}$$

and that we could therefore comprehend the process by which sugars are prepared in the plants them-

[1] *Ber. d. chem. Gesell.* 3, 66 (1870).

[2] *Ann. Chem.* (Liebig), 120, 295 (1861).

selves, since formaldehyde might readily be produced by the reduction of carbon dioxide by chlorophyll.

From this time on until Kiliani's demonstration of the actual constitution of several sugars, no work of very great importance was accomplished. Fittig, in a theoretical paper[1] now difficult of access, proposed that the sugars were derivatives of hexane, C_6H_{14}, viz. aldehydes of the various possible isomeric hexa-oxyhexanes. This view was the subject of much discussion. Whereas no serious objections were raised to Fittig's formula for glucose, as the aldehyde of normal mannite (identical with Baeyer's formula for methylenitane), Krusemann[2] showed that fructose also gave mannite on reduction, and therefore also contained a normal chain. Zincke[3] found that many ketone-alcohols gave oxyacids of the same number of carbon atoms when oxidized, and transferred this idea to the sugars, assigning the following formulæ:

Glucose: $CH_2OH-CHOH-CHOH-CHOH-CO-CH_2OH$

Fructose: $CH_2OH-CHOH-CHOH-CO-CHOH-CH_2OH$

They differ in the position given the carbonyl group. This view was adopted by Victor Meyer,[4] because the sugars did not give a certain aldehyde reaction, viz. a red color with a solution of fuchsine decolorized

---

[1] *Über die Constitution der sogenannten Kohlenhydrate* (Tübingen, 1871).

[2] *Ber. d. chem. Gesell.* 9, 1465 (1876).

[3] *l.c.* 13, 641 (1880); *Ann. Chem.* (Liebig), 216, 318 (1883).

[4] *Ber. d. chem. Gesell.* 13, 2343 (1880); Schmidt, *ibid.* 14, 1850 (1881).

ᘁ

with sulfurous acid.[1] The preparation of oximes[2] of the sugars by the action of hydroxylamine, showed that they must be either aldehydes or ketones.

The aldehyde formula for glucose was greatly strengthened by the fact that on oxidation it furnished acids of the same number of carbon atoms (gluconic and saccharic acids), though, as we have seen from Zincke's work, this did not preclude a ketone structure. In 1880, Kiliani[3] showed that fructose gave oxidation products containing less than six atoms of carbon; this, in conjunction with the formation of mannite upon reduction, proved fructose to be a *ketone* with normal chain. Four years later, Kiliani and Kleemann[4] proved the presence of a normal chain in gluconic acid, the only previous demonstration of this fact being the oxidation of gluconic to saccharic acid:

$$HOOC-CHOH-CHOH-CHOH-CHOH-COOH$$

and the reduction of this latter to adipic acid:

$$COOH-CH_2-CH_2-CH_2-CH_2-COOH$$

The direct reduction of gluconic acid gave an oxycapronic acid:

$$CH_3-CH_2-CHOH-CH_2-CH_2-COOH$$

respectively its anhydride, caprolactone.

[1] This "fuchsine-sulfurous acid reaction" has since then always occupied a prominent place in the discussions on the sugar question, far more prominent than it deserves; for its mechanism is unknown, and it is given by some compounds which are not aldehydes. No diagnostic value is at present attached to this reaction.

[2] V. Meyer and Schulze, *Ber. d. chem. Gesell.* 17, 1554 (1884); Rischbieth, *ibid.* 20, 2673 (1887); Wohl, *ibid.* 24, 993 (1891).

[3] *Ann. Chem.* (Liebig), 205, 190 (1880).

[4] *Ber. d. chem. Gesell.* 17, 1296 (1884).

Kiliani[1] next showed that the product obtained by oxidation of galactose, galactonic acid, is a normal-chain acid isomeric with gluconic acid, for it also yields caprolactone upon reduction. This proved galactose to be an aldehyde of normal hexa-oxyhexane, just as glucose is.

The chief reaction whereby Kiliani was enabled to establish the constitution of many sugars consisted in the application of a well-known and simple synthetic process. Aldehydes, as well as ketones, add hydrocyanic acid to form cyanhydrines,

$$>C{=}O + HCN = >C{<}^{OH}_{CN}$$

and these *oxynitriles* are readily saponified to acids. Now, if the sugars are aldehydes or ketones, they should add hydrocyanic acid, and the structure of the acids formed from these will show what the constitution of the sugar is. Thus glucose[2] was found to give a cyanhydrine; the acid prepared from this, when reduced, yielded *normal heptylic* acid :

1. $CH_2OH-CHOH-CHOH-CHOH-CHOH-CHO$

2. $CH_2OH-CHOH-CHOH-CHOH-CHOH-CH{<}^{OH}_{CN}$

3. $CH_2OH-CHOH-CHOH-CHOH-CHOH-CHOH-COOH$

4. $CH_3-CH_2-CH_2-CH_2-CH_2-CH_2-COOH$

This proved glucose to be the *aldehyde of mannite ;* it cannot be a ketone. Fructose,[3] on the other hand,

---

[1] *Ber. d. chem. Gesell.* **18**, 1555 (1885).

[2] *Ibid.* **19**, 1128 (1886).

[3] *Ibid.* **23**, 449 (1890); Düll, *ibid.* **24**, 348 (1891).

gave *methyl-butyl-acetic* acid, which can be explained only as follows :

1. $CH_2OH-CHOH-CHOH-CHOH-CO-CH_2OH$

2. $CH_2OH-CHOH-CHOH-CHOH-C {<}{^{OH}_{CN}} -CH_2OH$

3. $CH_2OH-CHOH-CHOH-CHOH-C {<}{^{OH}_{COOH}} -CH_2OH$

4. $CH_3-CH_2-CH_2-CH_2-CH {<}{^{CH_3}_{COOH}}$

Fructose is thus a ketone of mannite, with its carbonyl in β position.

Kiliani[1] extended his work to arabinose, a sugar obtained from cherry gum, to which the formula $C_6H_{12}O_6$ had been given. His results were startling. In the first place, he found that arabinose furnished *tetraoxyvalerianic* acid when oxidized; *i.e.* an acid with five atoms of carbon. The hydrocyanic acid addition product of arabinose treated as above gave a lactone $C_6H_{10}O_6$, which upon reduction was converted into caprolactone resp. caproic acid. Arabinose must therefore be an aldehyde, as it is a normal derivative, and contains only five atoms of carbon in the molecule; its formula is $C_5H_{10}O_5$. Upon a reduction it yielded a new alcohol, *arabite*, of the composition $C_5H_{12}O_5$. To-day, when we know sugars containing from three to nine carbon atoms, this seems of no great import. But as late as ten years ago, until this discovery of Kiliani's, it was axiomatic among chemists that all carbohydrates contained six atoms of carbon or some multiple thereof.

[1] *Ber. d. chem. Gesell.* 19, 3029 (1886); 20, 339 (1887); 20, 133 (1887).

These memorable researches of Kiliani would seem to have been sufficient for all purposes of constitutive and synthetic progress. Glucose was shown to be the aldehyde of mannite, galactose the aldehyde of an isomere of mannite, fructose to be a ketone derived from mannite, containing its carbonyl in β position, arabinose to be the aldehyde of a new penta-acid alcohol with normal chain. Truly remarkable results for a few years' work. But why did not Kiliani continue in the path he had followed with such success ? Why was it left for another to reap the harvest ?

Because of almost insurmountable experimental difficulties, requiring skill and patience and insight such as are seldom united in one individual. Even when pure, sugars crystallize with difficulty ; when mixtures are present, when other organic and inorganic materials are formed or introduced during the reactions, the task of isolating a sugar by ordinary means is hopeless. The man who discovered the magic touchstone, and who possessed the necessary skill and patience and insight, was Emil Fischer ; his wizard's wand was *phenylhydrazine*. Fischer's researches on the sugars date from 1887, and are laid down in innumerable separate articles. Fortunately their author has furnished two elaborate *résumés* [1] of his work, to which the reader is referred, and from which the following brief sketch is taken.

As is well known, phenylhydrazine reacts readily with carbonyl groups :

$$>CO + H_2N-NHC_6H_5 = >C=N-NHC_6H_5 + H_2O$$

[1] *Ber. d. chem. Gesell.* **23**, 2114 (1890); **27**, 3189 (1894).

A condensation of this base with the sugars, which we have found to contain the aldehyde or ketone grouping, is therefore to be expected. Fischer found, however, that the reaction is not simple in this case. For although the first step proceeds normally

$$CH_2OH-CHOH-CHOH-CHOH-\overset{*}{C}HOH-CH=N-NHC_6H_5$$

to form a *hydrazone*, this hydrazone is usually easily soluble, and escapes detection ; and, furthermore, it undergoes a very peculiar oxidation when warmed with excess of the hydrazine base. The alcohol group marked with a * in the above formula gives up its hydrogen atoms to pass into the carbonyl group ; this, then, reacts with more phenylhydrazine quite normally, producing an *osazone :*

$$CH_2OH-CHOH-CHOH-CHOH-C-CH=N-NHC_6H_5$$
$$\overset{\|}{N-NHC_6H_5}$$

The hydrogen atoms which leave the sugar reduce a molecule of phenylhydrazine to aniline and ammonia ; the complete reaction thus requires three molecules of the base to one of sugar.

These osazones possess the very desirable property of being quite insoluble in water, and easily crystallizable from other solvents. Thus they readily serve to detect the presence of sugars, and to identify them by their characteristic melting points. It was by means of this reaction that Fischer was able to isolate and identify his synthetic sugars. It is a very curious fact, however, which will be entered into in detail farther on, that several sugars give *identical* osazones ; thus, glucose, fructose, and man-

nose give one and the same derivative, viz. glucosazone (called thus because first obtained from glucose).

The reconversion of the osazones into their sugars
is a difficult task, which has succeeded but imper-
fectly. For though it is possible to split off the
phenylhydrazine group, there results invariably a
ketose,[1] whether the original sugar be aldose or ketose.
The steps of the reaction are as follows: First, concentrated hydrochloric acid replaces the hydrazone
groups with oxygen in a normal manner:

$$CH_2OH-CHOH-CHOH-CHOH-C-CH=N-NHC_6H_5 + 2 H_2O$$
$$\overset{\parallel}{N}-NHC_6H_5$$
$$= CH_2OH-CHOH-CHOH-CHOH-CO-CHO + 2 C_6H_5NH-NH_2$$

The aldehyde-ketone formed is called an *osone;* no
osones have yet been isolated in a sufficiently pure
state to prove the formula assigned them, but their
reactions render it extremely probable that the assumption is correct. Secondly, reduction of the
osones leads to ketoses, by a simple reaction.

In order to regenerate an aldose from an osazone,
it is necessary to reduce the ketose first obtained to
the corresponding alcohol, which then upon oxidation gives an aldose.[2] It is easily seen that the
method of recovering sugars from their phenylhydrazine compounds is far from satisfactory; it is re-
sorted to only in case of dire necessity.

The goal which Fischer set for himself was the

---

[1] Sugars containing an aldehyde group are called *aldoses, e.g.*
glucose; those containing a ketone group are called *ketoses, e.g.*
fructose.

[2] With some ketose as well.

synthetical preparation of the sugars, a problem almost as old as the science. Let us follow his progress, after a short survey of the previous literature of the subject.

The formation of sugars by oxidation of the poly-hydroxy-alcohols had been first observed by Carlet,[1] who found that dulcite behaved thus; the reaction was then studied by Gorup-Besanez[2] and by Dafert,[3] who showed that a mixture of sugars was obtained here. Van Deen[4] showed the formation of a sugar-like substance by the oxidation of glycerine, a fact rediscovered by Grimaux[5] much later. Fischer was able to greatly extend this chapter by proving: first, that the oxidation of dulcite gave fructose and a new sugar, mannose; second, that Van Deen's reaction gave a real sugar, glycerose, $C_3H_6O_3$, which, however, is a mixture of the two possible compounds, glycerine aldehyde and dioxyacetone:

$$CH_2OH-CHOH-CHO \quad \text{and} \quad CH_2OH-CO-CH_2OH$$

and third, that erythrite, the alcohol with four atoms of carbon, also gave a sugar, erythrose, $C_4H_8O_4$. Mannose was soon thereafter found in large quantities as a plant product in the so-called reserve-cellulose.

The polymerization of formaldehyde to methyl-enitane by Butlerow has already been referred to. This was the first real synthesis of a sugar, though

[1] *Jahresbericht*, 1860, 250.

[2] *Ann. Chem.* (Liebig), 118, 257 (1861).

[3] *Ber. d. chem. Gesell.* 17, 227 (1884).

[4] *Jahresbericht*, 1863, 501.

[5] *Compt. rend.* 104, 1276 (1887).

the proof of this came much later.   O. Loew[1] (1886) repeated Butlerow's work, and regarded the sugar he obtained as different from methylenitane; he called it *formose*.   Later[2] he succeeded in preparing a sugar which was fermentable, whereas formose and methylenitane were indifferent toward yeast; this compound he called *methose.*

By this time Fischer had already begun his attempts to synthesize a natural sugar.   Acroleïne di-bromide is converted into a sugarlike product by alkalies (best barium hydrate):

$$2\,C_3H_4OBr_2 + 2\,Ba(OH)_2 = C_6H_{12}O_6 + 2\,BaBr_2$$

The main product consisted of two sugars, which were named *α-* and *β-acrose*.   Fischer found that *α*-acrose was present in small quantity in methylenitane and formose, and in larger quantity in methose, to which it imparted the property of fermentability. What made *α*-acrose especially interesting, however, was the remarkable resemblance between *its* osazone and *glucosazone;* they differed only with regard to optical activity, *α*-acrosazone being inactive.

The evident conclusion to draw from this fact was that *α*-acrose represented the inactive modification of either glucose or fructose, — a conclusion borne out by subsequent investigation.   The chief difficulty, of course, consisted in procuring sufficient material for the many operations required.   The best method of preparing *α*-acrose was found to consist in the polymerization of glycerose, which we have seen to

L

[1] *J. prakt. Chem.* [2] **33**, 321 (1886).
[2] *Ber. d. chem. Gesell.* **22**, 475 (1889).

be a mixture of glyceric aldehyde and dioxyacetone; a simple "aldol condensation" suffices to give a sugar of the same structure as fructose :

$$CH_2OH-CHOH-CHO + CH_2OH-CO-CH_2OH =$$
$$CH_2OH-CHOH-CHOH-CHOH-CO-CH_2OH$$

The yield is small, as a number of other products are formed at the same time. The conversion of α-acrosazone into its sugar, α-acrose, was carried out by means of the above-described process (p. 122). α-Acrose is a syrup which ferments with yeast, gives levulinic acid when heated with hydrochloric acid, and is reduced by sodium-amalgam to a new hexaoxyalcohol. This was called α-acrite, and closely resembles mannite, save that it is optically inactive.

But by the time a kilo of glycerine has been transformed into α-acrite, the mass has shrunk to 0.2 gramme. Small hope of progress by this road. A series of fortunate circumstances, however, supplied the needed links in the chain.

Mannose, the aldehyde of mannite, may be readily oxidized to the corresponding *mannonic* acid. Owing to the difficulty with which all the compounds of the sugar group crystallize, this acid cannot be directly isolated; but phenylhydrazine proved its efficacy here by forming a well-defined *hydrazide*[1] of the acid :

$$CH_2OH-CHOH-CHOH-CHOH-CHOH-COOH + C_6H_5NH-NH_2 =$$
$$CH_2OH-CHOH-CHOH-CHOH-CHOH-CO-HN-NHC_6H_5 + H_2O$$

The acid is easily regenerated from this hydrazide, and obtained in the form of its lactone $C_6H_{10}O_6$. It

---

[1] Just as ammonia forms an *amide*, or aniline an *anilide*.

will be remembered that the lactone obtained by Kiliani from the cyanhydrine of arabinose also possessed this composition. A comparison of the two compounds gave the remarkable result that the acids were *identical* in every respect save their optical activity; Fischer's lactone turned the plane to the *right*, Kiliani's to the *left*. In aqueous solution they combined to form a third, *inactive*, substance.

We have here a case perfectly similar to that of dextro- and lævo-tartaric acid. In order to transfer these conditions to mannose itself, we need but to convert the lactones into their corresponding sugars. Now it is well known that the reduction of an acid to its aldehyde or alcohol is a very dubious operation. Fischer was therefore surprised and delighted to find that the *lactones* are reduced to aldehydes with ease and despatch on being treated with sodium amalgam in cold *acid* solution. This reaction, simple though it is, has been the chief factor in the wonderful development of the sugar group. Mannonic acid is thus readily reconverted into d-mannose;[1] Kiliani's arabinose-carboxylic acid into the isomeric l-mannose; the inactive acid into i-mannose.

These sugars may be reduced to their alcohols, giving three optically different *mannites*. Here is where the sections of the synthetic chain are forged together; for i-mannite is identical with the syn-

---

[1] The prefixes *d-*, *l-*, and *i-* denote the genetic relations of these compounds, and not the actual direction in which they rotate the plane of polarized light. All compounds derived from ordinary (d-)glucose are therefore d-compounds, no matter what the nature of their optical activity is. Thus, d-glucosazone rotates to the left; the sugar regenerated from it, fructose, also rotates to the left; but *genetically* it is *d*-fructose.

thetic α-acrite, and α-acrose is nothing else than i-fructose. There remains only to pass from the inactive series to the active in order to complete the synthesis of natural sugars.

For this purpose we possess two methods, discovered by Pasteur[1] during his classic research on racemic acid : partial fermentation, and fractional crystallization of salts. Only the former method is applicable to the sugars. α-Acrose is rapidly attacked by yeast ; the solution is no longer inactive, but turns the polarized plane to the right — it contains l-fructose.[2] Inactive mannose behaves similarly ; the dextro modification is destroyed, the lævo remains. The yeast thus feasts upon the forms it is accustomed to, leaving us the comparatively uninteresting optical antimeres[3] of the desired compounds.

Fischer sought refuge in chemical methods. i-Mannite was oxidized first to i-mannose, then to i-mannonic acid. The fractional crystallization of the strychnine or morphine salts of this last furnished the two active forms of mannonic acid, which could then be reduced to the corresponding sugars, d- and l-mannose. d-Mannose then leads to d-fructose by way of its osazone, since mannose and fructose give the same phenylhydrazine derivative.

Glucose and mannose also give the same osazone. Their difference must therefore be due to the asym-

---

[1] *Ann. chim. Phys.* (1848–1854).

[2] See note on opposite page.

[3] This word seems to be better adapted for general use than the customary "antipodes." Its meaning is quite clear, and it is used throughout this book to express optical as well as geometric isomerism.

metry of that carbon atom which in the osazone is
united to the *second* phenylhydrazine group, since
identity does not result until this asymmetry is
destroyed :

$$CH_2OH-CHOH-CHOH-CHOH-\overset{.}{C}HOH-CHO$$

It is a matter of general experience that when substances containing an asymmetric carbon atom are
heated, rearrangement takes place so as to form the
corresponding antimeres. *E.g.* tartaric acid is converted into racemic acid, and racemic into mesotar-
taric acid.   Since glucose and mannose differ in the
spatial grouping around just *one* carbon atom, a
similar readjustment might be expected.   The sugars
themselves are too unstable for such treatment, but
their acids are easily convertible — best when heated
with quinoline or pyridine.   Thus gluconic acid is
partly converted into mannonic acid when heated to
140° ; mannonic acid is similarly transformed into
gluconic acid, a condition of equilibrium being
reached.   As d-gluconic acid is easily reduced to
d-glucose, a complete synthesis of grape-sugar has
been effected.   Chart I. will show this in detail, together with the synthesis of other derivatives of
mannite.

Fischer was able to go farther than this, however;
for the continuation of the reactions discovered by
Kiliani and himself has led to the synthesis of sugars
containing more than six atoms of carbon.   First of
all, Fischer found that the addition of hydrocyanic
acid to the sugars always gives *two* stereo-isomeric
acids; thus arabinose yields l-*gluconic* acid along
with the l-mannonic acid discovered by Kiliani, and

thereby becomes a welcome source of experimental material. By means of the cyanhydrine reaction on the hexoses, and reduction of the resulting acids as described, sugars were obtained from mannose and

CHART I

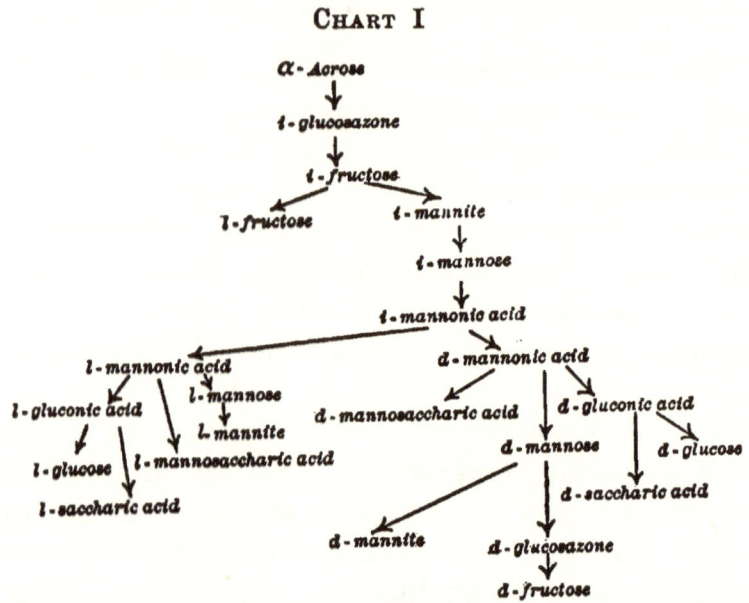

glucose containing seven carbon atoms. These *heptoses*, manno-heptose and gluco-heptose, by the same methods then gave corresponding *octoses* and *nonoses*, the latter having the formula:

$$CH_2OH-CHOH-CHOH-CHOH-CHOH-CHOH-CHOH-CHOH-CHO$$

Were it not that by the time these nonoses are reached, the supplies of material have dwindled down to microscopic dimensions, we should probably have sugars with ten and twenty carbon atoms in a chain; though whether the sugars would be any sweeter, we have no means of knowing.

These were the main results reached by Fischer at the time of his lecture before the " Chemische Gesellschaft" (1890).   The mannite group lay completely unravelled before the world.   Since then, the dulcite and sorbite families have been the object of care and attention, with considerable success.   The most important problem, however, consisted in the determination of the actual space configuration belonging to the various sugars.   No less than sixteen isomeric hexoses are possible, according to the theory of Le Bel and van t'Hoff, which possess the plane-structure of glucose.   Here, too, success has waited upon the efforts of the pioneer in the field, for no less than eleven of the possible sixteen forms are known to-day.

It is not the object of this chapter to give a complete catalogue of all the compounds of the sugar series, nor on the other hand to deduce and prove the phenomena of stereochemistry; we can therefore only develop the configurations of the hexoses, and their immediate derivatives, as being the most important, as far as needed for this purpose.   The method of formulating the asymmetry of carbon atoms is quite simple.   If we imagine the carbon tetrahedron :

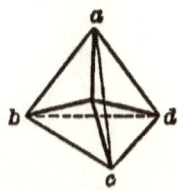

projected upon the plane of the paper, the four valences will be spread out as follows :

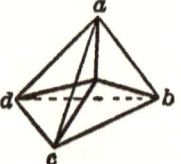

In the case of the optical antimer of the first tetra-hedron:

the projection will look like this:

the arrows indicating the *difference in direction* of rotation. Regarded as *plane* formulæ, these projections are of course identical; considered as plane-pictures of the *solid* formulæ, they are different. Corresponding projections are obtained with two or more asymmetric carbon atoms:

As a matter of economy in writing, as well as of convenience in reading, it has become customary to represent the asymmetric carbon atoms in such formulæ by the intersection of the valence lines (just as in aromatic compounds the benzene nucleus is represented by a hexagon):

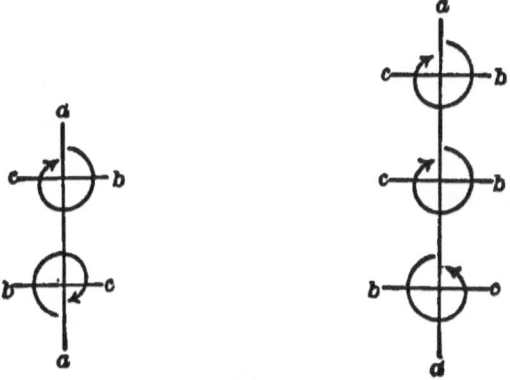

The hexoses contain *four* asymmetric carbon atoms:

$$CH_2OH-CH(OH)-CH(OH)-CH(OH)-CH(OH)-CHO;$$

likewise the monobasic acids formed from them; and in the subsequent development the formula of the sugar will be used for the acid as well. It will be readily seen that by no possibility can the compounds become optically inactive by inner compensation as long as we deal with the sugars themselves and their monobasic acids; for the opposite ends of the molecule are *unlike:* CH_2OH on the one hand, CHO resp. COOH on the other. Thus even a perfectly symmetrical arrangement of the other substituents cannot bring about a *complete* symmetry of the molecule:

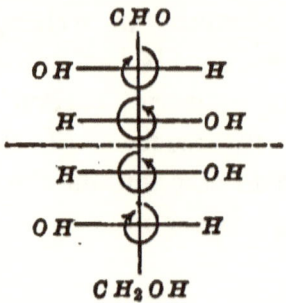

The case is different with the *dibasic* acids and with the alcohols. Here the ends of the molecule are *identical*, and therefore when the remaining substituents are symmetrically grouped:

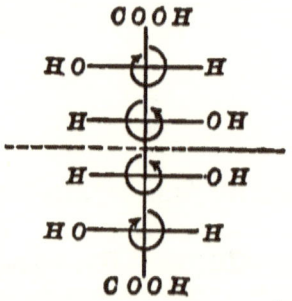

the compound becomes optically *inactive by intramolecular compensation*, as each half of the molecule is the mirrored image of the other.

It will not be an easy task for the reader to follow Fischer's deductions from here on; continual cross-references, which make the line of reasoning appear confused, are necessary in order to bring out the genetic connections. But with a little patience, and study of the formulæ, the difficulties will vanish. The formulæ will be numbered in the order in which taken up, and reference to them will always be by this number; for convenience' sake they are given in consecutive order on Chart II.

Two different sugars, glucose and gulose, give *saccharic acid;* therefore all the stereochemical possibilities for this acid are contained in the following formulæ:

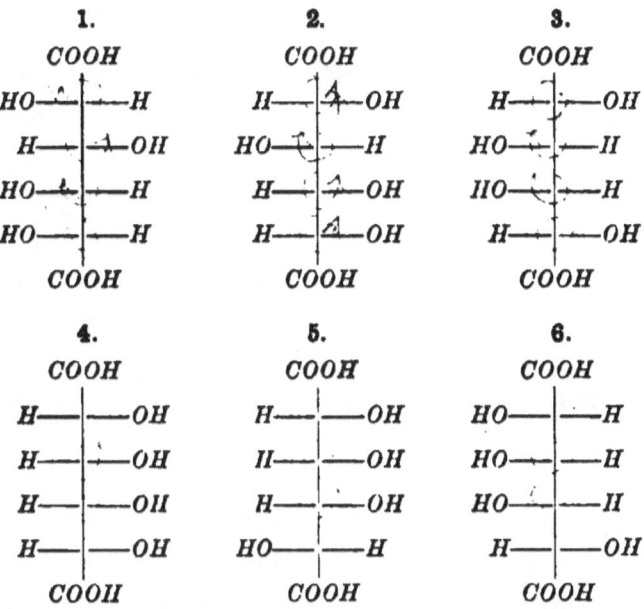

Nos. **3** and **4** are at once excluded, as they represent *inactive* compounds; and the following considerations show that **5** and **6** are also unsuitable. Mannosaccharic acid, which is formed by oxidation of mannose, stands in the same relation to mannose as saccharic acid does to glucose. Now we have seen that the two sugars differ only with respect to the asymmetry of the carbon atom next to the aldehyde group (for they give the same osazone). Therefore if saccharic acid is either **5** or **6**, mannosaccharic acid must be either **4** or **3**. But mannosaccharic acid is *not* in-

CHART II

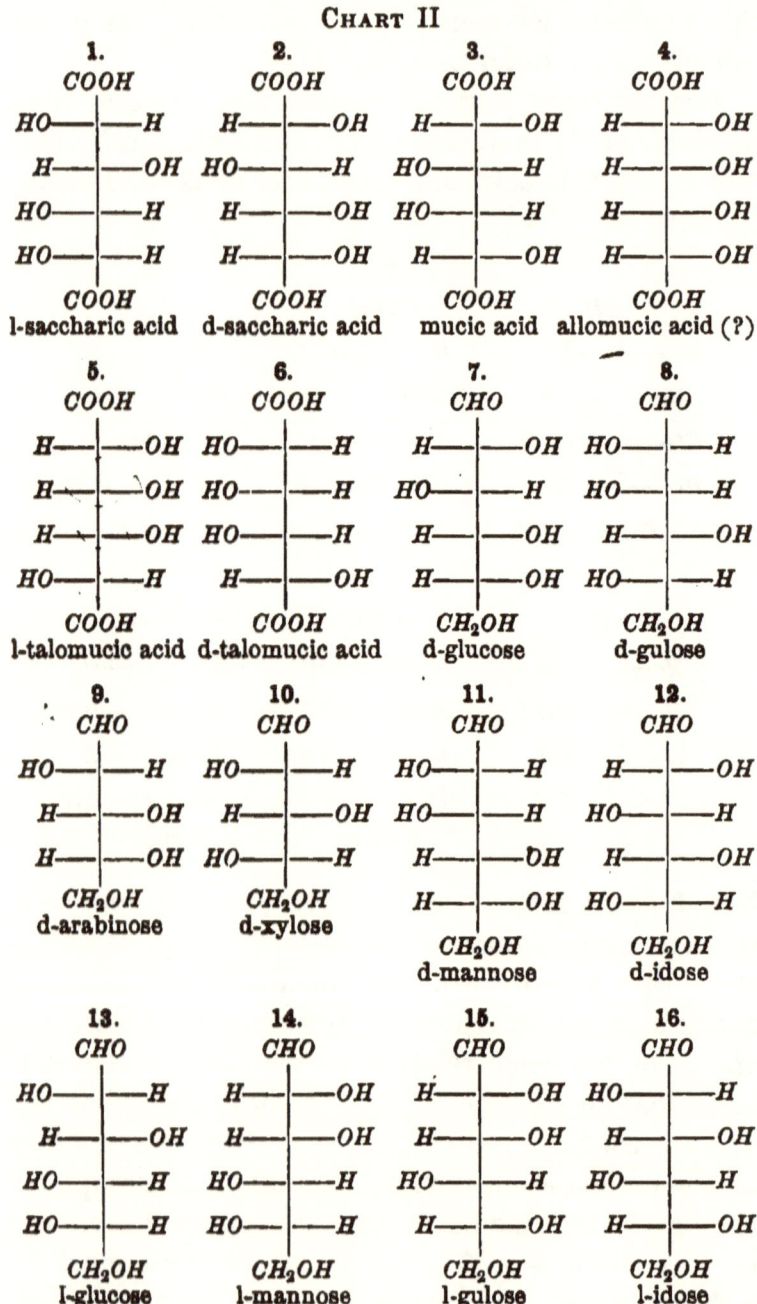

1. l-saccharic acid
2. d-saccharic acid
3. mucic acid
4. allomucic acid (?)
5. l-talomucic acid
6. d-talomucic acid
7. d-glucose
8. d-gulose
9. d-arabinose
10. d-xylose
11. d-mannose
12. d-idose
13. l-glucose
14. l-mannose
15. l-gulose
16. l-idose

CHART II (*continued*)

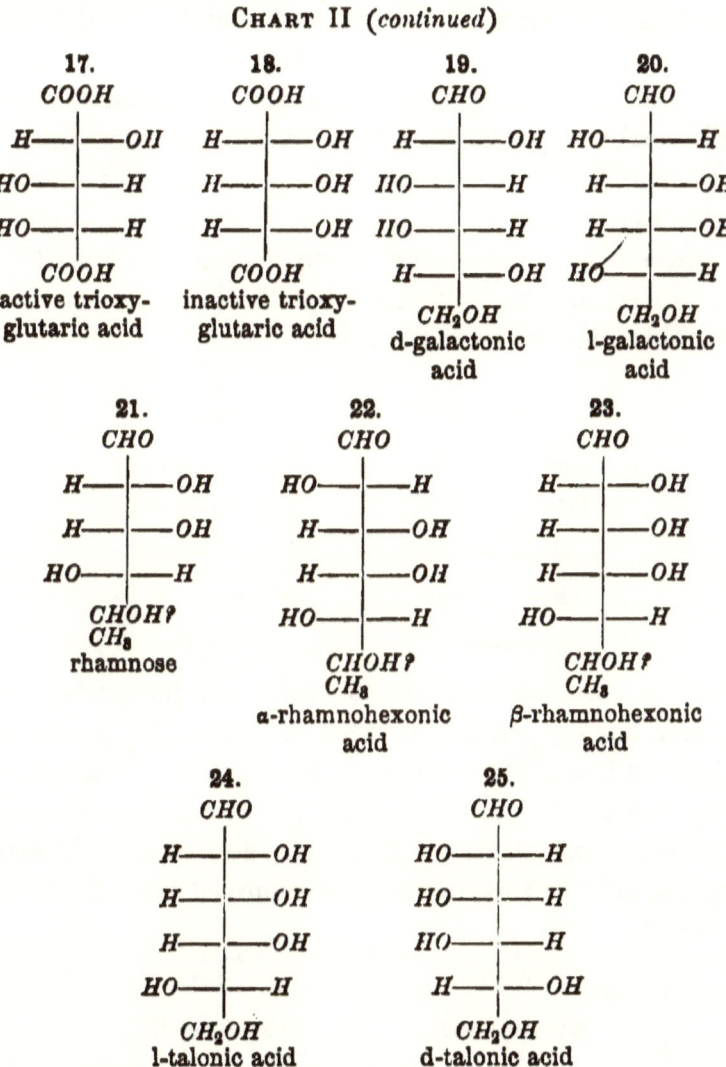

17.
COOH
H——OH
HO——H
HO——H
COOH
active trioxy-
glutaric acid

18.
COOH
H——OH
H——OH
H——OH
COOH
inactive trioxy-
glutaric acid

19.
CHO
H——OH
HO——H
HO——H
H——OH
CH₂OH
d-galactonic
acid

20.
CHO
HO——H
H——OH
H——OH
HO——H
CH₂OH
l-galactonic
acid

21.
CHO
H——OH
H——OH
HO——H
CHOH?
CH₃
rhamnose

22.
CHO
HO——H
H——OH
H——OH
HO——H
CHOH?
CH₃
α-rhamnohexonic
acid

23.
CHO
H——OH
H——OH
H——OH
HO——H
CHOH?
CH₃
β-rhamnohexonic
acid

24.
CHO
H——OH
H——OH
H——OH
HO——H
CH₂OH
l-talonic acid

25.
CHO
HO——H
HO——H
HO——H
H——OH
CH₂OH
d-talonic acid

active, therefore does not possess formula **4** or **3**, and consequently saccharic acid cannot be **5** or **6**. Thus only **1** and **2** remain for consideration. These formulæ represent merely the dextro and lævo modification of the same compound, as they are to each other as object and mirrored image. Since we have no means of deciding at present which *actual* order of arrangement corresponds to right and to left, Fischer *arbitrarily* selected **2** for d-saccharic acid. This arbitrary choice, however, defines once for all the relative arrangements of all compounds connected with saccharic acid.

d-Saccharic acid thus being **2**, and being further produced by the oxidation of both d-glucose and d-gulose, these latter must be :

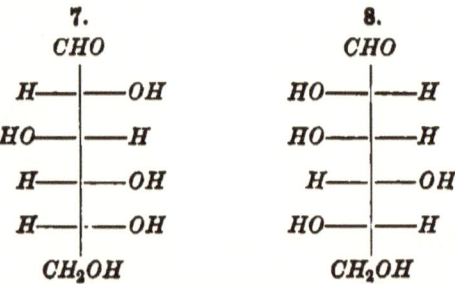

Which formula belongs to each, can be settled by remembering their connections with the *pentoses.* Glucose is derived from arabinose, gulose from xylose (by the cyanhydrine reaction). If the asymmetric carbon atom resulting from the addition of hydrocyanic acid to these sugars be removed from **7** and **8**, we get :

Only one of these, **10**, can yield an *inactive dicarboxylic acid.* Now, as xylose gives inactive trioxyglutaric acid, it possesses this formula (accurately speaking, it belongs not to *ordinary* (lævo) xylose, but to its optical antimer). **9** is then the formula of d-arabinose. From these facts we conclude that d-glucose is **7**, d-gulose **8**.

We are now able to derive the formula of mannose. As we have seen, this sugar differs from glucose merely by the asymmetry of the carbon atom nearest the aldehyde group, since the osazones of these sugars are identical. By reversing the arrangement of substituents on this carbon atom, we get:

11.

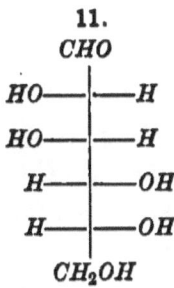

CHO
HO——|——H
HO——|——H
H——|——OH
H——|——OH
CH₂OH

as the stereochemical picture of d-mannose.

Fischer obtained a new sugar from gulose in the same manner in which glucose is converted into mannose, viz. by oxidizing it to gulonic acid, heating with pyridine, and reducing the lactone of the new acid thus obtained, *idonic acid*, to the corresponding sugar, *idose*. The formula of this new sugar is thus derived from that of gulose (**8**), by the process just applied to mannose:

12.

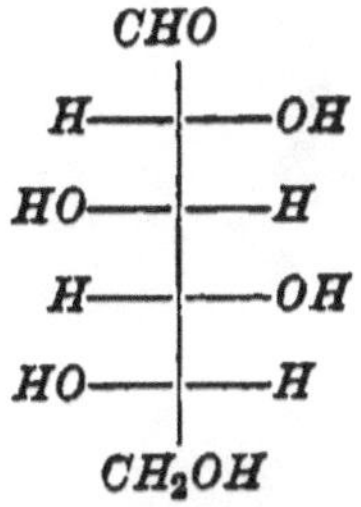

CHO
H——|——OH
HO——|——H
H——|——OH
HO——|——H
CH₂OH

Having deduced the configurations of glucose (**7**), mannose (**11**), gulose (**8**), and idose (**12**) in their dextro modifications, we are able to give those of their optical antimeres offhand:

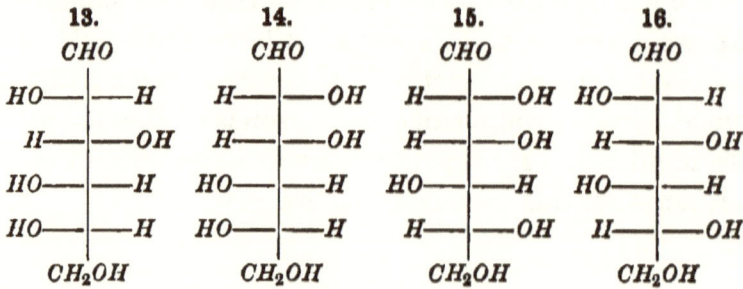

Eight of the possible sixteen normal hexa-aldoses are therefore determined with an accuracy dependent only upon the legitimacy of the fundamental stereochemical assumptions. As another method of procedure, starting from xylose, leads to exactly the same configurations as those just given, the reliability of these deductions of Fischer's is greatly strengthened.

The dulcite family, including galactose and mucic acid, must be diagnosed without reference to the preceding group (the mannite family). *Mucic acid* is optically *inactive*, and therefore possesses one of the following formulæ:

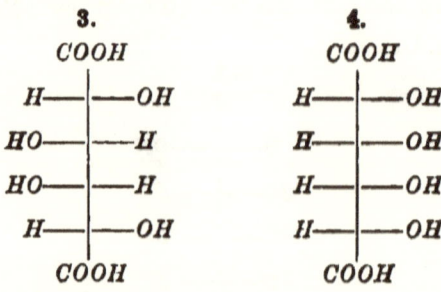

The chain of reasoning which leads to **3** is rather complicated. Rhamnose, which is methyl-pentose :

$$CH_3-CHOH-CHOH-CHOH-CHOH-CHO$$

furnishes an *active trioxyglutaric acid* upon oxidation (one carbon atom being completely split off). Now α-rhamnohexonic acid, which is obtained from rhamnose by the cyanhydrine reaction, similarly splits off a carbon atom when oxidized and yields mucic acid, which is *inactive*. The asymmetric carbon atom introduced by addition of hydrocyanic acid must therefore have converted an *active* system of three carbon atoms into an inactive one of four. Only **3** meets this requirement; for of the trioxyglutaric acids which might be obtained from the two formulæ :

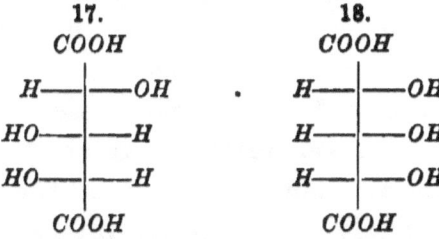

17 is active, and 18 inactive. Mucic acid therefore possesses configuration **3**.

D- and l-galactose both give mucic acid upon oxidation. They therefore have the following optically antimeric formulæ :

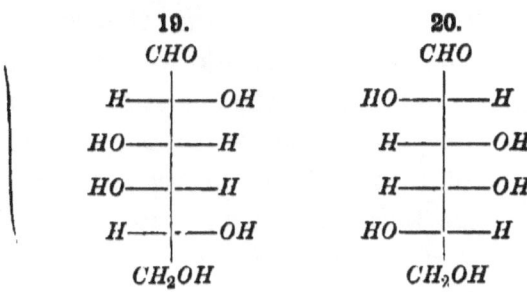

but which one of these belongs to each, cannot be determined directly, since mucic acid is inactive by intramolecular compensation. No transitions from the mannite to the dulcite group are known, or the matter would be very much simplified. A few facts must be first taken into account. If d-galactonic acid is heated with pyridine, it is in part converted into a new acid, *talonic* acid; and these two acids are, of course, antimeric with respect to the carbon atom attached to the carboxyl. d-Talonic acid reduces to a new sugar, d-*talose*, and, on oxidation, goes over into a dibasic acid, called d-*talomucic acid*. Mucic and talomucic acids are therefore also antimeric in the just-mentioned sense. Finally, α-rhamnohexonic acid, which we have seen to furnish mucic acid when oxidized, is converted into *its* antimer, β-rhamnohexonic acid, by heating with pyridine, and this β-rhamnohexonic acid passes over into a talomucic acid upon further oxidation — the acid being the *optical* (lævo) antimer of the d-acid obtained (indirectly) from galactonic acid. These facts suffice to develop the configuration of d- and l-galactose. They show first that when rhamnose (**21**, Chart II.) and its derivatives are oxidized, the *methyl* group is split off; for the α- and β-rhamnohexonic acids are antimeric with reference to the first carbon atom (counting from the carboxyl); mucic and talomucic acids are likewise antimeric in the same sense. Now when the rhamnohexonic acids are oxidized, they lose a carbon atom; but as they yield mucic and talomucic acids respectively, containing the *same* antimeric carbon atom, it follows that the carbon atom split off does not come from the original carboxyl group.   Therefore it comes from the other end of the chain, viz. the methyl group.   We have seen that rhamnose gives an *active* trioxyglutaric acid when oxidized, to which formula **17** was ascribed:

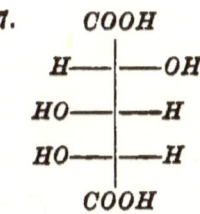

17.   COOH

Following out the genetic relations detailed above, we get the series of configurations belonging to the group:

22.   COOH   23.   COOH

22 belongs to α-rhamnohexonic acid, because this gives mucic acid (**3**); **23** thus remains for the antimeric β-acid.   Oxidation of **23**, with elimination of the methyl, gives for l-talomucic acid formula **5**; d-talomucic acid therefore possesses the optically antimeric configuration **6**:

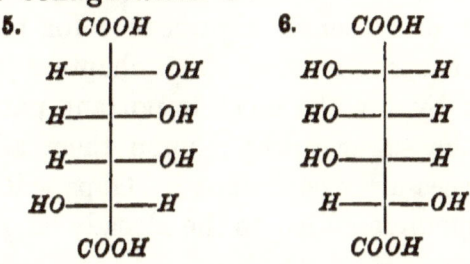

5.   COOH   6.   COOH

By reduction of these acids, d- and l-talonic acids are to be obtained; and their relations to the rhamnohexonic acids prove that the lower carboxyl group is reduced, giving:

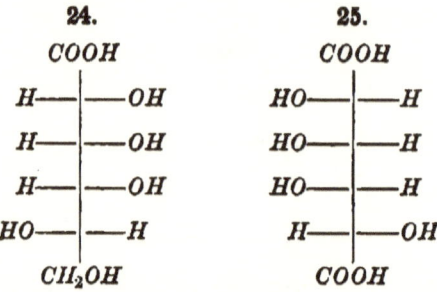

l-Talonic acid and l-talose (**24**) are still unknown. d-Galactonic and d-talonic (**25**) acids are antimeric with respect to a single carbon atom; therefore d-galactonic acid possesses formula 19, and l-galactonic acid **20**:

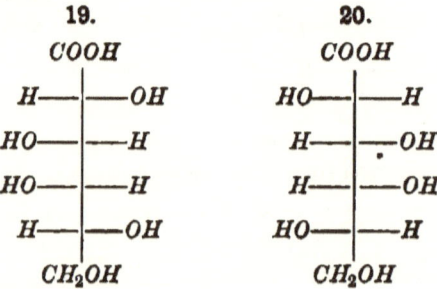

Eleven isomeric hexo-aldoses thus exist to verify the predictions of stereochemical theory; truly an achievement of no small significance for the legitimacy of such speculations. The chapters just completed form by far the most important part of the history of the sugar group; upon them all future work will necessarily be founded. Gaps will be filled in, new compounds added to the already lengthy list;

valuable data will be made available for physico-chemical study; and the chemistry of assimilation is now a possibility. This last direction seems the most promising, and many interesting discoveries have been made by Fischer, who is eagerly following up the trail. Thus, he has found that only sugars containing 3, 6, or 9 carbon atoms are fermentable; further, that a certain arrangement of the hydroxyl groups favors fermentation; again, that yeast attacks only one of the optically antimeric forms of a given sugar; that enzymes are selective in their action on sugar derivatives; and more of similar importance and interest.[1]

In order to simplify the nomenclature and the registration of compounds of the sugar group, Fischer has proposed the following system: The international nomenclature is to be employed in naming the compound, whereas the stereochemical arrangement of the hydroxyl groups is to be denoted by the + and − signs. The former indicates the right, the latter the left, of the median line. Thus, to take two examples, glucose would be called:

Hexose + − + +   or   *hexanpentolal* + − + +

while d-saccharic acid is catalogued as:

Hexantetroldiacid + − + +   or   − − + −

As a much needed complement to Fischer's synthetic methods, A. Wohl[2] has given us a reaction for "building down" the sugars, *i.e.* for successively

---

[1] A full review of the history of the sugars during 1896 is given by W. E. Stone, *Amer. Chem. J.* 19, 608 (1897).

[2] *Ber. d. chem. Gesell.* 26, 730 (1893).

removing carbon atoms from the molecule. Wohl availed himself of the fact that oximes easily lose water to form nitriles:

$$-CH:NOH = -CN + H_2O$$

Glucosoxime undergoes this reaction when heated with acetic anhydride (the hydroxyl groups being of course acetylated at the same time); the (acetylated) gluconitrile thus obtained:

$$CH_2OH-CHOH-CHOH-CHOH-CH\Big\langle{{OH}\atop{CN}}$$

loses hydrocyanic acid on treatment with ammoniacal silver nitrate solution, to pass into (acetyl) pentose:

$$CH_2OH-CHOH-CHOH-CHOH-CHO$$

The pentose itself, finally, is obtained by successive action of ammonia and dilute acids upon the acetyl body.

In this manner Wohl obtained d-arabinose from d-glucose; this new sugar, mixed with ordinary l-arabinose, furnished the inactive, racemic, arabinose. The method promises to be extremely fruitful,[1] and has very recently been employed by Fischer.[2] Starting with rhamnose, he obtained a methyl-tetrose whose configuration must be:

$$\begin{array}{c} CHO \\ H-|-OH \\ HO-|-H \\ CHOH? \\ CH_3 \end{array}$$

---

[1] Since writing the above, Wohl has applied this method to galactose, with the result of confirming Fischer's formula. *Ber. d. chem. Gesell.* **30**, 3101 (1897).

[2] *Ibid.* **29**, 1378 (1896).

Oxidation of this leads to d-tartaric acid ; now, as in all cases heretofore observed, the oxidation of methyl-aldose leads to elimination of the methyl group, a similar course of reaction is to be predicated here.   Therefore, d-tartaric acid has the following configuration :

$$
\begin{array}{c}
\text{COOH} \\
\text{H}-|-\text{OH} \\
\text{HO}-|-\text{H} \\
\text{COOH}
\end{array}
$$

Similarly important results are to be expected shortly.

There remains to be discussed a general formula for the sugars known as " Tollens's [1] ethylene-oxide formula."   Owing chiefly to the fact that sugars do not give the fuchsine-sulfurous acid test for aldehydes, as well as that sugars differ from ordinary aldehydes in being stable in the air, it has been suggested by Tollens, and contended by others,[2] that the aldoses are not really aldehydes, but inner anhydrides of hepta-oxy compounds, similar to ethylene oxide :

$$CH_2OH-CHOH-CH-CHOH-CHOH-CH-OH$$
$$O$$

Such compounds would readily give the normal reactions of aldehydes with hydroxylamine and phenylhydrazine ; in fact, by the simple assumption of addition of water, whenever needed, the oxide formula could readily pass into the ordinary one :

---

[1] *Ber. d. chem. Gesell.* **16**, 923 (1883).

[2] Sorokin, *J. prakt. Chem.* **37**, 312 (1888); Skraup, *Monatsh.* **10**, 401 (1889).

$$
O\begin{cases} \begin{array}{c} CHOH \\ | \\ CHOH \\ | \\ CHOH \\ | \\ CH \\ | \\ CHOH \\ | \\ CH_2OH \end{array} \end{cases} + H_2O \longrightarrow \begin{array}{c} CH{<}^{OH}_{OH} \\ | \\ CHOH \\ | \\ CHOH \\ | \\ CHOH \\ | \\ CHOH \\ | \\ CH_2OH \end{array} \longrightarrow \begin{array}{c} CHO \\ | \\ CHOH \\ | \\ CHOH \\ | \\ CHOH \\ | \\ CHOH \\ | \\ CH_2OH \end{array} + H_2O
$$

Very recent facts seem to make this view plausible,[1] though at present we must wait for something more than plausibility.

A subject intimately connected with this formula of Tollens is that of "multirotation." Many sugars have a different specific rotation in their freshly prepared solutions than when the solutions have been standing for some time. Usually the initial rotation is greater than the final; such is the case with glucose, where the ratio is as 2 to 1, or with fructose, whose ratio is as 10 to 9, or with xylose, whose original rotation is four times greater than its final. The reverse phenomenon has also been observed; maltose rotates *less* in its freshly prepared solutions by the ratio of 8 to 9, rhamnose as 1 to 4. These curious facts have been adduced in *support* of the "oxide" formula of these sugars, as well as in refutation thereof. It is readily seen that on the assumption just illustrated (that water may add on to either the aldehyde or the ethylene-oxide group) we can "prove" both formulæ. Fischer has suggested that perhaps the aldehyde structure is to be preferred;

---

[1] *E.g.* the existence of isomeric glucoses. Tanret, *Compt. rend.* **120**, 1060 (1894).

for the sugar could dissolve as aldehyde, and gradually add water to form the hepta-oxy-alcohol. Recent investigations[1] show, however, that the molecular weight of the dissolved sugar in its initial and final conditions of rotation is one and the same, which fact precludes the addition hypothesis. It has therefore been proposed[2] that the two sugar solutions differ merely in containing first the aldehyde, and then the oxide structure ; the two forms must necessarily have different rotatory powers, since the oxide contains a new asymmetric carbon atom.   Certain reactions, as yet undeveloped, appear to indicate that the oxide is the final structure; such being the ready conversion of glucose into fructose and mannose by dilute alkalies.[3] These reactions must be studied farther before any conclusions can be drawn from them. The real cause of multirotation is still a mystery.

The chemistry of the compound sugars, such as cane and milk sugars, maltose, etc., is in a very unsatisfactory condition. It is well known that these carbohydrates are easily hydrolyzed by acids into simpler ones ; cane sugar yields glucose and fructose, milk sugar gives glucose and galactose, maltose furnishes two molecules of glucose. This process is known as *inversion*, because in the case first studied (cane sugar) the direction of rotation is changed from right to left. This reaction of the compound sugars has given them the name "polysaccharides,"

---

[1] Trey, *Ztschr. physik. Chem.* **18**, 193 (1895); **22**, 424 (1897).

[2] Lippmann, *Chemie der Zuckerarten* (1895); Lobry de Bruyn, *Ber. d. chem. Gesell.* **28**, 3081 (1895).

[3] Lobry de Bruyn and van Ekenstein, *l.c.* 3078.

or *bioses*, *trioses*, etc., according to the number of simpler sugars they form. The polysaccharides are divided into two distinct groups: the members of one give all the characteristic reactions with phenyl hydrazine, and therefore contain the aldehyde or ketone radical; the members of the second group are indifferent to this reagent, and thus do not contain carbonyl. The first group includes milk sugar (lactobiose), maltose (maltobiose), and isomaltose[1] (isomaltobiose); these sugars show all the typical reactions of the monosaccharides, reducing Fehling's solution, adding hydrocyanic acid, forming monobasic acids when oxidized, etc. The second group contains cane sugar (saccharobiose), besides other not well-known substances (trehalose, etc.). Various formulæ[2] have been proposed for these different sugars, but as they rest more upon suppositions than facts, they need not be considered here for the present.

Fischer is engaged in the study of still another branch of the sugar family: the so-called glucosides. It is a very well-known fact that many substances occur in nature combined with glucose, in rather unstable compounds, *e.g.* amygdaline, salicine, arbutine, digitaline, etc. The glucosides are readily hydrolyzed, by acids, alkalies, or enzymes; amygdaline gives glucose, benzaldehyde, and hydrocyanic acid; salicine gives glucose and saligenine (o-oxybenzyl alcohol). Occasionally other sugars occur in this

---

[1] A "synthetic" biose, obtained by Fischer [*Ber. d. chem. Gesell.* **23**, 3687 (1890)]. It is formed by action of hydrochloric acid on glucose. It has since been found in malt and artificial dextrine.

[2] *Ber. d. chem. Gesell.* **26**, 2405 (1893).

manner. The first synthesis of a glucoside was effected by Michael,[1] who prepared salicine from salicylic aldehyde, by treating its potassium salt with "acetochlorhydrose" (obtained by action of acetyl chloride on glucose; it possesses the formula : $C_6H_7O . Cl(O . C_2H_3O)_4$). Many others have been artificially prepared since then, without any particular addition to our knowledge of the manner in which the glucose group is attached to the rest of the molecule. Fischer is attacking this problem by the investigation of the simpler compounds of the series. This study has not led to very tangible results, so far, though it has been rich in experimental facts. All the sugars seem to give glucosides with various alcohols,[2] in the presence of a *small* amount[3] of hydrochloric acid. The reaction proceeds in two steps; the first consists in the formation of an *acetal:*

$$CH_2OH - CHOH - CHOH - CHOH - CHOH - CHO + 2\,CH_3OH =$$

$$CH_2OH - CHOH - CHOH - CHOH - CHOH - CH\underset{OCH_3}{\overset{OCH_3}{<}} + H_2O$$

The acetal then loses one molecule of alcohol, and passes into the glucoside. The nomenclature of these compounds is simple; their names are formed by taking that of the sugar as well as that of the alcohol concerned; *e.g.* methyl glucoside, propyl mannoside, butyl galactoside. · It has been much easier to name these substances than to ascribe a formula to them. *Two* isomeres[4] are usually formed;

[1] *Amer. Chem. J.* 1, 309 (1879).
[2] *Ber. d. chem. Gesell.* 26, 2401 (1893).
[3] *Ibid.* 28, 1145 (1895).
[4] *Ibid.* 27, 2985 (1894).

this is easily comprehensible, since by the process a new asymmetric carbon atom has been produced. Two formulæ have been proposed:

$$
\text{I.} \quad O
\begin{cases}
CH-OCH_3 \\
\;\;|\\
CH-OH \\
\;\;|\\
CH-OH \\
CH \\
CH-OH \\
CH_2OH
\end{cases}
\qquad
\text{II.} \quad
O<
\begin{matrix}
CH-OCH_3 \\
CH \\
CH-OH \\
CH-OH \\
CH-OH \\
CH_2OH
\end{matrix}
$$

the one by Fischer, the other by Franchimont[1] and by Marchlewski.[2] They differ with regard to the carbon atom which is drawn into the reaction; in all other respects they agree with the observed isomerism of the glucosides, with their indifference toward alkalies, Fehling's solution, and phenylhydrazine, and their extreme sensitiveness toward acids (which split them into their components). Time must decide between these formulæ. Fischer has found that the sugars readily give *mercaptals*[3] with the thio-alcohols, of the general structure:

$$
C_5H_{12}O_5 . C<
\begin{matrix}
SCH_3 \\
SCH_3
\end{matrix}
$$

upon the investigation of which he bases great hopes. The sugars also condense with one and with two molecules of various ketones, but the constitution of these compounds is not yet known. Here, too, the future will bring us needed light.

In closing this sketch of the sugar group, it may not be amiss to refer to a few compounds which were

[1] *Recueil trav. chim.* **12**, 312 (1893).

[2] *Ber. d. chem. Gesell.* **26**, 2928 (1893); **28**, 1622 (1895).

[3] *Ibid.* **27**, 673 (1894).

once included in the family, but have since been found to be interlopers. Chief among these is inosite, $C_6H_{12}O_6$. Maquenne[1] has shown this substance to be hexaoxyhexahydrobenzene:

$$\begin{array}{c} H(OH) \\ H(OH) \quad H(OH) \\ \bigcirc \\ H(OH) \quad H(OH) \\ H(OH) \end{array}$$

Three modifications of it are known, as well as many methyl-substitution products (pinite, dambonite, quebrachite), all of which are sugar-like bodies. The simplest member of the sugary hydrated benzenes was recently prepared by Baeyer,[2] who has named it *chinite;* it is p-dioxyhexamethylene,

$$\begin{array}{c} H(OH) \\ H_2 \quad H_2 \\ \bigcirc \\ H_2 \quad H_2 \\ H(OH) \end{array}$$

and resembles inosite in its general properties.

The sugars were the despair of the earlier chemists because of their complex reactions and uninviting properties. Scarcely twelve years have elapsed since the commencement of Kiliani's and Fischer's researches; this brief period has seen the establishment of a chapter of science whose completeness is probably unparalleled. Many gaps remain to be filled; with the advance of our knowledge of facts

[1] *Ann. chim. phys.* Series 6, 12, 87 (1887).
[2] *Ann. Chem.* (Liebig), 278, 88 (1894).

has come the realization of deeper problems awaiting their solution. No one seems to feel this more keenly[1] than the man who has led us thus far; but no one has less reason to despair of success in these more difficult fields than Emil Fischer.

[1] Cf. *Ber. d. chem. Gesell.* **27**, 3190 (1894).

CHAPTER VII

THE ISOMERISM OF MALEÏC AND FUMARIC ACIDS

THE problems of isomerism have always possessed a peculiar interest and importance in the history of chemistry. Aside from the curiosity such questions invariably arouse, the existence of isomeric substances serves as a touchstone for testing the worth of our theories. When Liebig, in the dawn of organic chemistry, discovered that cyanuric acid was absolutely identical in composition with cyanic acid, his first impression was that a huge mistake had been made somewhere. Chemical theories demanded that identity of composition should amount to absolute identity. When the fact of isomerism became clearly established, it was manifestly necessary to expand the theoretical horizon. An explanation for the case cited was found in the difference of molecular weights possessed by the two compounds: polymerism at least partially accounted for isomerism. When, subsequently, isomerism appeared among compounds of equal molecular weight and comparatively similar behavior, the next step in the evolution of chemical theory was the establishment of our modern structural system. In this way, the existence of, *e.g.*, four butyl alcohols finds its satisfactory justification. And when, finally, the limitations of structural theory failed to account for two isomeric alpha-lactic acids, and for four dihydroxysuccinic acids, the unravelling of the isomerism involved gave us the theory of the asymmetric carbon atom in all its beauty and completeness.

What wonder, then, that the existence of two substances certainly not polymeric, containing no asymmetric carbon atom, and apparently structurally identical, seemed little short of a miracle. Yet the case of isomerism presented by maleïc and fumaric acids could hardly be considered under the head of any other known phenomena. And as might have been expected by historical analogy, the solution of this problem has given us a new fundamental conception of molecular statics.

When malic acid is subjected to dry distillation,[1] the chief products are fumaric acid and the anhydride of maleïc acid. The latter substance is easily converted into maleïc acid itself. Both acids have the same composition, $C_4H_4O_4$, and are diabasic; their formulæ may therefore be partially resolved as follows:

$$C_2H_2(COOH)_2$$

In properties, derivatives, etc., there is the widest possible difference between these two substances. Fumaric acid does not melt, but volatilizes above 200°; maleïc acid melts at 130°, and boils at 160°, at the same time decomposing into its anhydride and water. The former acid is difficultly soluble in water, the latter easily. Barium fumarate is soluble, barium malëate is insoluble. The dimethyl ester of the former is a solid, of the latter a liquid. Fumaric acid crystallizes in irregular needles, maleïc acid in

---

[1] Pelouze, *Ann. Chem.* (Liebig), **11**, 263 (1834).

rhombic prisms.   Evidently, it would not be easy to mistake the one substance for the other.

But the most curious fact regarding the relation of the two acids is the ease with which they are converted into each other.   Thus, all attempts to convert fumaric acid into its anhydride, whether by heating or any other process, yield as sole product the anhydride of *maleïc* acid.   On the other hand, if maleïc acid be maintained at a temperature slightly above its melting point, it slowly solidifies as fumaric acid. A similar though much more rapid transition is effected by concentrated halhydric acids, hydriodic acid acting much more rapidly than hydrobromic or hydrochloric acids.   Maleïc ester passes into fumaric ester when warmed with iodine.

We owe most of our systematic knowledge of these transitions to Skraup,[1] who has shown that these processes do not take place quantitatively.   In all cases, other substances (chiefly addition products) are formed at the same time.   Thus, in the above reaction with hydrobromic acid, monobromsuccinic is also to be found in the resulting product.   Skraup also discovered that substances which by themselves have no effect upon maleïc acid, partly convert it into fumaric acid in the presence of atoms upon which they can react.   Thus, hydrogen sulphide does not act upon maleïc acid ; but if copper malëate be decomposed by the gas, a considerable quantity of fumaric acid is produced.   Evidently, it must be a very subtle and unusual variety of isomerism which unites these curious acids.

[1] *Monatsh.* **12**, 107 (1891).

The earliest explanations of the phenomenon, naturally enough, regarded it as a case of polymerism. Thus Liebig,[1] in his classic treatise on the constitution of organic acids, compares this case to that of cyanic and cyanuric acids. Erlenmeyer[2] remarked that acids might be regarded as oxy-aldehydes:

$$-\overset{H}{\underset{}{C}}=O \qquad\qquad -\overset{OH}{\underset{}{C}}=O$$

and that therefore polymerization is to be expected among acids, aldehydes being peculiarly prone to this form of condensation. This idea was finally disposed of by Anschütz. Kekulé and Anschütz had shown that when maleïc acid[3] is oxidized it is converted into meso-tartaric acid; fumaric acid,[4] under the same conditions, gives racemic acid. Now racemic acid, being a mixture of dextro- and lævo-tartaric acids, was regarded as a bimolecular substance; and its relation to fumaric acid would justify the assumption that the latter is also bimolecular. Anschütz[5] showed that the esters of both racemic and fumaric acids are mono-molecular; this removed the sole basis of the polymeric hypothesis.

The question received considerable impetus, if not marked insight, from careful experimental investigation by Fittig. Fittig[6] found that both acids, when

---

[1] *Ann. Chem.* (Liebig), **26**, 168 (1838).

[2] *Ber. d. chem. Gesell.* **3**, 342 (1870); **19**, 1937 (1886). See also Markownikoff, *Ann. Chem.* (Liebig), **182**, 356 (1876).

[3] *Ber. d. chem. Gesell.* **13**, 2150 (1880).

[4] *Ibid.* **14**, 713 (1881).

[5] *Ann. Chem.* (Liebig), **239**, 161 (1887). A brief account of the earlier history of these substances is given in this article.

[6] *Ibid.* **188**, 98 (1877).

treated with nascent hydrogen, gave succinic acid. This proved beyond a doubt the presence of the following chain of carbon atoms in the molecule :

$$\text{HOOC} - \overset{|}{\underset{|}{\text{C}}} - \overset{|}{\underset{|}{\text{C}}} - \text{COOH}$$

leaving the two hydrogen atoms still to be accounted for. A second result of this investigation was the formation of two different dibrom-succinic acids, according to which of the two substances was treated with bromine. Each of these dibrom products was reduced to succinic acid, thus proving their relation to this latter substance.[1] Fittig's third result was the formation of one and the same mono-brom-succinic acid by addition of hydrobromic acid to either of the isomeres. The conclusion which Fittig draws is that since the only formulæ possible[2] for two dibrom-succinic acids are :

$$\text{I.} \quad \begin{array}{l} \text{CHBr} - \text{COOH} \\ | \\ \text{CHBr} - \text{COOH} \end{array} \quad \text{and} \quad \begin{array}{l} \text{CH}_2 - \text{COOH} \\ | \\ \text{CBr}_2 - \text{COOH} \end{array} \quad \text{II.}$$

the formulæ of maleïc and fumaric acids must be represented by :

$$\text{I.} \quad \begin{array}{l} \text{CH} - \text{COOH} \\ \| \\ \text{CH} - \text{COOH} \end{array} \quad \text{and} \quad \begin{array}{l} \text{CH}_2 - \text{COOH} \\ | \\ = \text{C} - \text{COOH} \end{array} \quad \text{II.}$$

Of these two structures he assigns II. to maleïc acid.

Fittig's view was confirmed by Beilstein,[3] who stated that the dibrom-succinic acid obtained from

[1] They are also formed when succinic acid is brominated directly.

[2] It must be remembered that this statement antedates the stereochemical epoch.

[3] Beilstein and Wiegand, *Ber. d. chem. Gesell.* 15, 1499 (1882).

maleïc acid, when treated with moist silver oxide, yielded pyruvic acid. This reaction is easily explained by the following formulæ:

$$\underset{CBr_2COOH}{\overset{CH_2COOH}{|}} \longrightarrow \underset{C(OH)_2COOH}{\overset{CH_2COOH}{|}} \longrightarrow \underset{C(OH)_2COOH}{\overset{CH_3}{|}} \longrightarrow \underset{CO-COOH}{\overset{CH_3}{|}}$$

But a reinvestigation of Beilstein's results by V. Meyer and Demuth[1] showed the former to have been in error; no pyruvic acid is obtained in this reaction.

It is interesting to consider the condition of the problem at this stage. The experimental results of Fittig, V. Meyer, and Demuth left not a shadow of doubt that both maleïc and fumaric acids were unsaturated succinic acids of the structure:

$$\underset{CH-COOH}{\overset{CH-COOH}{\|}}$$

Structural chemistry, it would appear, had exhausted its resources; and the opinion was freely expressed that help lay only in the direction of spatial extension of our formulæ. But how? and when? The answer came almost immediately upon the question. It was another illustration of the inventive energy of necessity.

But while stereochemical speculations were gathering strength and material, a final attempt was made to save the day for structural theories. Anschütz,[2] in an elaborate experimental and critical treatise, advanced a new formula for maleïc acid. He showed

[1] *Ber. d. chem. Gesell.* **21**, 264 (1888). The substance is really brom-fumaric acid.
[2] *Ann. Chem.* (Liebig), **239**, 161 (1887).

that Fittig's old formula for this substance might as well have been assigned to fumaric acid; and that, secondly, it could belong to neither, owing to the relations existing between these acids and racemic and meso-tartaric acids.   Anschütz's new formulation is as follows:

$$CH\diagup\begin{matrix}OH\\OH\end{matrix}\atop\|\atop CH\diagup\begin{matrix}O\\O\end{matrix}$$

The chief argument in support of this structure of maleïc acid appears to be that no really binding objection to it can be stated; but it assumes an equivalent for the carboxyl groups in an undoubted acid, a proposition hardly to be admitted without debate.   However, the formula never took on formidable proportions; even before it was offered to the public, the long-expected, generalized solution of the difficulty appeared.

Wislicenus,[1] in a brief monograph, summed up the situation in a manner which has since received almost universal approval, and become a fundamental principle of organic chemistry.   It is true, there have been objections and criticisms, which we shall in part consider later; but there is little doubt of the permanent value of Wislicenus's theory of "geometrical isomerism."

Wislicenus gives us a development of the earlier standpoint of stereochemistry.   It will be remembered that the fundamental conception of van t'Hoff and Le Bel lay in the assumption of a tetrahedral

[1] "Über die räumliche Anordnung der Atome," etc.   Trans. Saxon Acad. of Sciences (Math. physical section), 14 (1887).

form of the carbon atom, or at least of a tetrahedral spatial distribution of the four valences of the carbon atom. This assumption led to the asymmetry of any molecule containing a carbon atom attached to four different groups of atoms (cf. pp. 130 ff.). This molecular asymmetry conditioned the existence of two spatially similar (or enantiomorphic) forms of one and the same structural formula. And it will be further remembered that such space isomeres differed inappreciably in all their physical and chemical properties,[1] save one: the behavior toward polarized light. It is plain that the isomerism of maleïc and fumaric acids is not of this order; and, moreover, the substances do not contain asymmetric carbon atoms at all.

Van t'Hoff[2] had summarized his views into three propositions:

1. The radicals of a compound C-*abcd* cannot exchange places spontaneously.

2. Carbon atoms united by one unit of affinity (bond or valence) rotate freely around their common axis.

3. Freedom of rotation is inhibited by double or triple linkings.

Van t'Hoff himself had drawn certain conclusions from these theorems, but except in so far as they applied to optical isomerism they had escaped gen-

---

[1] It must be remembered, however, that the accumulation of asymmetric carbon atoms in a molecule may produce extreme diversity in chemical properties. Witness the chemistry of the sugars.

[2] *Stereochemistry*, translated by J. E. Marsh. (Clarendon Press, 1891.)

eral notice. In particular, he had clearly enunciated the formulæ[1] now accepted for our two acids; but the extension of the principle to a large number of cases, and the addition of three other theorems, is due to Wislicenus.

If we imagine two carbon atoms linked by a single bond, we get the following stereochemical picture of their union; and three forms are possible, according to whether the rotation of substituents in the order *a-b-c-d* (where *d* represents the point of linking) is the same in both atoms, or different:

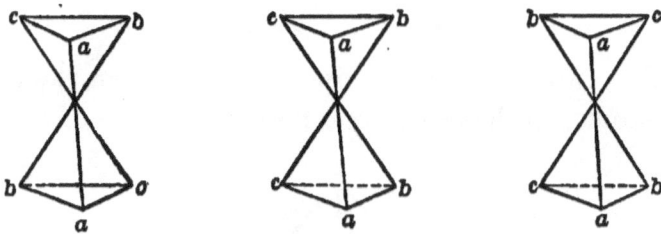

In order to save space, it has become customary to abbreviate these configurations thus:

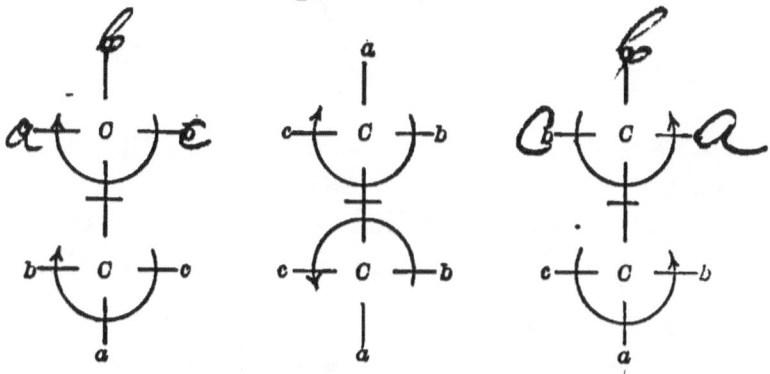

When the carbon atoms are doubly linked, the following appearance is presented; and here two forms

---

[1] *La chimie dans l' Espace* (1877).

are possible, according to whether the two *a* atoms are adjacent or not:

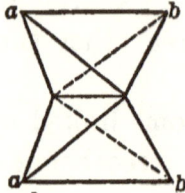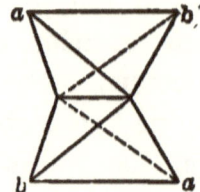

Here again it is convenient to save space by attaching spatial significance to our plane formulæ:

$$a - C - b \qquad a - C - b$$
$$\parallel \qquad\qquad \parallel$$
$$a - C - b \qquad b - C - a$$

Finally, in the case of a triple bond between the carbon atoms, we have but one possibility:

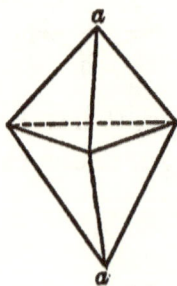

and stereo-isomerism is inconceivable in terms of our theory.

On this basis, van t'Hoff and Wislicenus assign the following formulæ to maleïc and fumaric acids:

$$H - C - COOH \qquad\qquad H - C - COOH$$
$$\parallel \qquad\qquad\qquad\qquad \parallel$$
$$H - C - COOH \qquad\qquad HOOC - C - H$$

Maleïc acid   Fumaric acid

This choice is determined by the fact that maleïc acid readily forms an anhydride, whereas fumaric acid forms none; it is quite reasonable to attribute the ease of reaction in the one case to proximity of the carboxyl groups, its absence in the other to the distance between them.

In order to determine the configuration of other substances whose isomerism is caused by a double linking, Wislicenus has set up three subsidiary theses; these have the further function of explaining the behavior of such isomeric substances under varying conditions. The first of these propositions runs thus: When a compound containing a triple linking passes by addition into a compound with a double linking, we can *a priori* determine the position taken up by the entering constituents. This is explained by the following diagram:

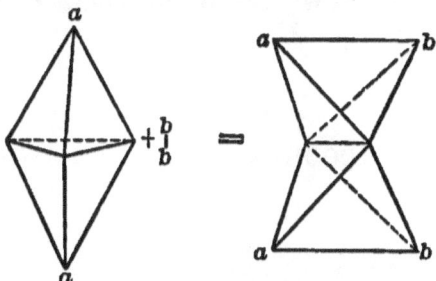

The two atoms *bb* take up the *cis*-position, as it is called.[1] This form of reaction, applied in the preparation of known substances, ought to settle, once and for all, the spatial relation of their radicals. But unfortunately, fact and theory do not yet go hand in hand, for Michael[2] has adduced a large

---

[1] Because they are on the same side of the plane of union. The opposite case is called the *trans*-position.

[2] *J. prakt. Chem.* [2], **46**, 402 (1892); **51**, 517 (1895). Michael's work has been partly nullified by later investigations, but a large number of contradictory data still remain.

number of instances in which the theory is sadly deficient.

Wislicenus's second thesis is best illustrated by an example. To account for the conversion of maleïc into fumaric acid by means of a small quantity of hydrobromic acid, we have the following series of diagrams. First, the maleïc acid adds on a molecule of hydrobromic acid:

$$
\begin{array}{ccc}
\text{H}-\text{C}-\text{COOH} & & \text{H} \\
\quad\quad\parallel \quad\quad\quad + \text{HBr} = & & | \\
\text{H}-\text{C}-\text{COOH} & & \text{H}-\text{C}-\text{COOH} \\
& & | \\
& & \text{H}-\text{C}-\text{COOH} \\
& & | \\
& & \text{Br}
\end{array}
$$

forming monobrom-succinic acid. The two halves of the molecule now rotate on their common axis into this position:

$$
\begin{array}{ccccc}
\text{H} & & \text{H} & & \\
| & & | & & \\
\text{H}-\text{C}-\text{COOH} & = & \text{COOH}-\text{C}-\text{H} & = & \text{COOH}-\text{C}-\text{H} \\
| & & | & & \parallel \\
\text{H}-\text{C}-\text{COOH} & & \text{H}-\text{C}-\text{COOH} & & \text{HC}-\text{COOH} \\
| & & | & & +\text{HBr} \\
\text{Br} & & \text{Br} &
\end{array}
$$

and hydrobromic acid being split off,[1] there results fumaric acid. The hydrobromic acid thus acts like a ferment, a small quantity sufficing to convert an unlimited quantity of maleïc acid. It will be admitted that this explanation possesses much elegance; but it does not accord with reality, as

---

[1] If this rotation is repeated with the aid of a model, it will be noted that a *different* hydrogen atom separates than the one first added.

Anschütz,[1] Michael,[2] and Skraup[3] have shown.   If the explanation be true, then monobrom-succinic acid, the supposed intermediate product during the reaction, ought to lose hydrobromic acid easily and form fumaric acid.   As a matter of fact, it does not do so except under conditions by no means comparable to those of the former reaction.

The third of Wislicenus's theorems concerns itself with the rotation just mentioned.   Given a molecule:

$$
\begin{array}{c}
b \\
| \\
a - C - c \\
| \\
e - C - g \\
| \\
f
\end{array}
$$

there must be certain attractions between **a, b, c, e, f,** and **g** in such manner that *one* position of the two carbon atoms must be more stable than the others. If, for example, **a** possesses a very strong attraction for **f,** then the two atoms will rotate until **a** is as near as possible to **f** ; *i.e.* until **a** and **f** are superposed :

$$
\begin{array}{c}
b \\
| \\
a - C - c \\
| \\
f - C - e \\
| \\
g
\end{array}
$$

This more stable position is thus the "preferred" configuration of the molecule, and while the position

---

[1] *Ann. Chem.* (Liebig), **254**, 168 (1889).   See also Fittig, *ibid.* **259**, 30 (1890).

[2] *J. prakt. Chem.* [2], **38**, 21 (1888).

[3] *Monatsh.* **12**, 107 (1891).

may be disturbed by a number of outside agencies, the tendency to return to the preferred location must in the end predominate. It is for this reason that in the above case of transition to fumaric acid the monobromsuccinic acid molecule rotated as shown; the latter position is preferred.

Here again there are unfortunate contradictions to disturb and perplex us. It so happens that the radicals seem to be inconstant in their preferences; and while Wislicenus sets up one table of attractive selection, Baeyer and Bischoff,[1] who are concerning themselves with similar phenomena in other fields, arrive at entirely different relations. We cannot within the limits of this volume discuss situations as yet entirely unsettled, and must therefore pass over the arguments which make out radical A to be more "positive" or more "negative" than radical B — perhaps the remedy will come when the misleading electrical analogies[2] are discarded in favor of an unbiassed study of facts.[3]

Yet, in spite of the failure of Wislicenus to completely prove his standpoint, the conception of geometrical isomerism has been extremely valuable to the organic chemist. Even though many details are imperfect and contradictory, we have been enabled to grasp under a common head a vast number of facts for whose comprehension we had no guide. We must trust to time to remove the discrepancies or evolve the successor of our present theory.

As to the critics of Wislicenus's point of view, they

---

[1] Cf. the discussion in R. Meyer's *Jahrbuch*, 1891, 133.
[2] Cf. Michael's "positive-negative" hypothesis, p. 77.
[3] Cf. Lossen, *Ann Chem.* (Liebig), 300, 30 (1898).

must bide their time until they offer us something better. In so far as they merely adduce contradictions, we must wait for a thorough sifting of evidence and of theory. Only two arguments have been advanced against the theory as such. Anschütz[1] cannot see how so slight a difference in spatial arrangement suffices to produce such difference in properties.[2] However, the world is full of wonders, and this seems to be one of them. After all, it is not so strange that Kolbe[3] greeted the first edition of van t'Hoff's book as the outburst of a diseased imagination. Michael argues differently. To him there is an essential difference between double and single linking; Wislicenus — van t'Hoff would make us believe that a double union between carbon atoms is simply the sum of two single unions. Michael, with perfect consistence, therefore avoids the use of double bonds in his formulæ, indicating free valences by dots; *e.g.* his formulæ for maleïc (and fumaric) acid:

$$\dot{C}H-COOH$$
$$|$$
$$\underset{.}{C}H-COOH$$

As to the phenomenon of isomerism, Michael offers us no explanation; he calls it allo-isomerism, to indicate its difference from other kinds of isomerism. Michael further objects to Wislicenus's theory because of the discrepancies noted above.

The majority of chemists hold with Wislicenus

---

[1] *Ann. Chem.* (Liebig), **239**, 165 (1887).

[2] Cf. footnote, p. 161.

[3] *J. prakt. Chem.* [2], **15**, 473 (1877). This passage is quoted in the edition of Van t'Hoff's work noted above.

rather than with Michael. Indeed, we have no other choice, for the theory of geometric isomerism is at least tangible, and has been a valuable incentive to further investigation. However, it would appear that Michael's energetic opposition to Wislicenus's views might be mitigated by the force of one of Michael's own propositions. It is undoubtedly true that the mechanical interpretation given for the transition of maleïc into fumaric acid does not stand the test of experiment; may not the real reason be *the very difference between double and single linkings upon which Michael insists?* If a double bond is not supposed to merely open on one side to leave a single one, why count the experimental evidence that it does not do so against the fundamental view of the structure of the molecule?[1] It is much better to admit the weight of evidence in favor of the latter assumption, and admit with Michael that the mechanism of transition is as yet hidden from our knowledge. There is no cause for doubting that before long this mystery also will be disclosed to the eager eye of the investigator.

[1] Skraup, *Monatsh.* 12, 107 (1891), assumes a sort of internal resonance as the cause of transition. The oscillations produced by certain reactions are absorbed by the maleïc acid molecule, which is thereby thrown into sympathetic oscillation.

CHAPTER VIII

THE ISOMERISM OF THE OXIMES

THE crucial test of any scientific theory is its ability to rise to unforeseen emergencies. Witness the steady development of Dalton's atomic theory — in spite of the unparalleled expansion of all branches of chemistry and physics, not a single fact is to-day incompatible with the theorems of 1808 and 1811 (the date of Avogadro's hypothesis). The doctrine of the linking of carbon atoms is another example of this innate elasticity of a really fundamental theory; modern structural chemistry has not yet outgrown the foster-mother of its infancy. And so it has been with the van t'Hoff-Le Bel conception of the spatial distribution of carbon valencies. We have seen how easily this theory accounted for the complicated phenomena among the sugars; we have learned with what success it has been applied to the isomerism of unsaturated compounds. It was not to be expected that a hypothesis primarily intended to explain the coexistence of carbon compounds would be capable of direct application to the derivatives of other elements. Yet recent years have shown such to be the case, and thus a new field has been conquered for stereochemistry.

The element in question is nitrogen; the derivatives in point are the *oximes.* As is well known,

170

carbonyl compounds (aldehydes and ketones) act upon hydroxylamine [1] as follows:

$$\underset{(H)R}{\overset{R}{>}}C \lceil O + H_2 \rceil NOH = \underset{(H)R}{\overset{H}{>}}C = NOH + H_2O$$

The resulting compounds are called oximes (*ald*oximes or *ket*oximes). They possess the character both of bases and acids. As acids, they are soluble in alkalies; as bases, they form rather unstable salts, such as the following:

$$C_6H_5CH=NOH . HCl$$

They serve for the important purpose of separating and identifying their mother-substances, for these properties enable us to handle them with facility in the laboratory.

Among the numerous reactions of the oximes, but few need to be recounted in this place. In the first place, the oximes are readily broken down into their constituents by the action of concentrated acids:

$$>C=NOH + H_2O = >CO + H_2NOH$$

The action of acetic anhydride upon them is peculiar; ketoximes react normally to form acetates:

$$\underset{R}{\overset{R}{>}}C=NOH + (CH_3CO)_2O = \underset{R}{\overset{R}{>}}C=NO . COCH_3 + CH_3COOH$$

aldoximes, on the other hand, lose water and form nitriles:

$$\overset{\lceil H}{R-C=N \rceil OH} = R-C\equiv N + H_2O$$

[1] V. Meyer and various pupils, *Ber. d. chem. Gesell.* **15**, 1324, 1525, 2778 (1882); **16**, 170 (1883).

This latter reaction shows a close connection between aldoximes on the one hand, and acid amides on the other; in fact, these isomeric substances differ but slightly in structure :

$$R-CH=NOH \qquad\qquad R-CO-NH_2$$

But it must be frankly admitted at this stage that we have no valid proof of the structure just assigned to the oximes. At the time of their discovery, V. Meyer and Janny deduced it from the fact that when treated with benzyl chloride in alkaline solution acetoxime forms an O-ester :

$$(CH_3)_2C=NO . C_7H_7$$

But the history of acetoacetic ester has taught us how little faith is to be placed in reactions of this sort (cf. p. 73). And as a matter of fact, we shall soon see that isomeric N-esters are now known which have been prepared in a precisely similar manner :

$$R_2CH{\Large<}\!\!\begin{array}{c} NC_7H_7 \\ | \\ O \end{array}$$

However, except for a certain limited period, there has been no disposition upon the part of the chemical world to question V. Meyer's original structure of the oximes, as this best expresses the general behavior of these substances.

As has been mentioned above, the oximes are important because they serve to identify aldehydes and ketones (substances notoriously difficult to manage when not perfectly pure). We can therefore readily understand how the discovery of *isomeric* oximes immediately became an object of public concern; for the diagnostic value of these derivatives was impaired.

The substance benzil is a well-known diketone which is very easy to procure, and which therefore is frequently employed in the prosecution of characteristic ketone reactions. Thus, V. Meyer had prepared both the mono- and the di-oxime derived from it, early in the course of his oxime-research:

$$C_6H_5-C=O \qquad C_6H_5-C=O \qquad C_6H_5-C=NOH$$
$$C_6H_5-C=O \qquad C_6H_5-C=NOH \qquad C_6H_5-C=NOH$$

Now if these formulæ actually represent the structure of these substances, isomerism is hardly to be thought of; and Meyer was considerably astonished to find later that his collaborator, H. Goldschmidt, had run across a second modification of benzildioxime.[1] The press of other duties prevented an immediate examination of this remarkable phenomenon, and it was not until 1888 that the siege was begun.

In the meantime another instance of isomerism among the oximes had come to light. In 1886 Beckmann[2] had discovered the curious molecular rearrangement which bears his name (*Beckmann'-sche Umlagerung*). When oximes are treated with dehydrating agents (phosphorus pentachloride, acetyl chloride, fuming sulphuric acid, etc.), and then poured into water, they are rearranged into

---

[1] *Ber. d. chem. Gesell.* **16**, 2176 (1883).

[2] *l.c.* **19**, 988 (1886).

acid amides; thus, benzophenone oxime yields ben-
zanilide :

$$C_6H_5-C=(NOH)-C_6H_5 \longrightarrow C_6H_5-CO-NH-C_6H_5$$

This reaction serves to emphasize the close connection between oximes and amides. Now, in applying
this reaction to the oxime of benzaldehyde, Beckmann found that instead of producing the corresponding acid amide, benzaldoxime gave birth to a
new isomere.[1] Five possible structures suggest
themselves for these substances :

  I.   II.   III.   IV.   V.

Of these, I belongs to benzaldoxime itself; II is the
structure of benzamide; III the tautomeric form of
benzamide (cf. p. 84); V is the formula of formanilide; IV thus remains for *iso*benzaldoxime, as
the new substance was called.   This formula is
strengthened by the peculiar property of isobenzaldoxime of promptly passing back into ordinary
benzaldoxime; it will be seen that the simple shifting of a hydrogen atom suffices to convert one
structure into the other.

But while the existence of two benzaldoximes
offered no remarkable difficulties of interpretation, a
severe strain was put upon structural theories by the
two dioximes of benzil.   It is true, a number of

---

[1] The isomerizing agent first employed in this case was dilute
sulphuric acid. *Ber. d. chem. Gesell.* **20**, 2766 (1887).   We now
use hydrochloric acid in ether solution.

"blackboard formulæ" can be constructed to face any emergency; but in this particular instance such a method of escape was precluded by the painstaking demonstration given by V. Meyer and Auwers[1] of the *structure-identity of the oximes in question.* As these investigators propose a stereochemical solution of the problem, it was clearly their duty to absolutely preclude the ordinary structural variety of isomerism. This demonstration was based upon the following facts: In the first place, both oximes are still derivatives of benzil,[2] and therefore have not undergone the Beckmann rearrangement. Secondly, both contain two hydrogen atoms directly replaceable by acetyl. Thirdly, both yield the same oxidation product : [3]

$$C_6H_5C=N-O$$
$$C_6H_5C=N-O$$

And finally, the oximes possess the same molecular weight, and are therefore not polymeric. In passing, it may be noted that Meyer and Auwers here for the first time in the history of orgánic chemistry employed Raoult's cryoscopic methods to determine molecular weights — a method brought to perfection by Beckmann, also in connection with the oximes.

We may thus regard it as reasonably certain that the two dioximes possess the same structure :

$$C_6H_5-C=NOH$$
$$C_6H_5-C=NOH$$

---

[1] *Ber. d. chem. Gesell.* **21**, 784, 3510 (1888).
[2] As they can be reconverted into benzil.
[3] Viz. diphenylglyoxime hyperoxide.

and that the cause of isomerism is to be sought in the spatial arrangement of their atoms. But the application of van t'Hoff's laws (cf. p. 161) presented serious difficulties. In the first place, the substances contain no asymmetric carbon atom. In the second place, the *carbon* atoms are all united by single linkings; this permits of freedom of rotation, and therefore excludes isomerism. The argument followed by V. Meyer and Auwers is that since no asymmetric carbon atoms are present, and since the isomerism is an unavoidable fact, the theorem that single bonds do not prevent free rotation is subject to exceptions.

In particular, the oximes of benzil are such an exception. If we arrange the central carbon atoms of these substances according to the abbreviated formulæ of Wislicenus (p. 162), we get the following possible spatial configurations (in which the symbol $n-n$ represents one oximido group):

$$
\begin{array}{ccc}
\underset{n-C-C_6H_5}{\overset{n}{\diagup|}} & \underset{C_6H_5-C-n}{\overset{n}{|\diagdown}} & \overset{C_6H_5}{|} \\
\underset{\overset{\diagdown|}{n}}{n-C-C_6H_5} & \underset{\overset{\diagdown|}{n}}{n-C-C_6H_5} & n-C-n \\
& & \underset{\overset{\diagdown|}{n}}{n-C-C_6H_5}
\end{array}
$$

Instead of merely representing phases of rotation, we must now suppose the formulæ to be more or less stable molecular entities. Similar considerations apply to benzil monoxime, and Meyer and Auwers thus predict the existence of three modifications each of mono- and di-oxime. A further consequence of this new theory is that the isomerism of the two benzaldoximes cannot be traced to the same cause, since they contain only one available carbon atom.

Fortune favored the new theory ; for Meyer and Auwers soon found a second monoxime of benzil.[1] Almost immediately afterwards, the third modification of benzil dioxime[2] was discovered. And to clinch matters, benzophenone[3] yielded but *one* oxime, as predicted.

Meanwhile, the investigation of isobenzaldoxime was producing interesting results in Beckmann's hands.[4] The possibility of polymerism was excluded by the molecular weight of the new substance. The formula of a tautomeric benzamide (imidobenzoic acid):

$$C_6H_5C \overset{OH}{\underset{NH}{<}}$$

improbable as it was, lost caste altogether by the fact that isobenzaldoxime did not show the behavior of an imido-ether, and that but *one* benzyl group can be introduced. There remained only the structure already mentioned :

$$C_6H_5CH \overset{NH}{\underset{O}{<}}_{|}$$

This solution was emphasized by the results of alkylation. On treatment with benzyl chloride in alkaline solution, ordinary benzaldoxime[5] (usually distinguished as *alpha*-oxime) gave an *α*-benzyl ester,

---

[1] For theoretical reasons this was called γ-benziloxime. *Ber. d. chem. Gesell.* **22**, 537 (1889).

[2] *l.c.* **22**, 705 (1889).　　　　[3] *l.c.* **22**, 537 (1889).

[4] *l.c.* **22**, 429 (1889).

[5] *l.c.* **22**, 513, 1531 (1889).

a liquid, whereas iso-(β)benzaldoxime yielded a *solid* β-benzyl ester. A closer examination of these esters showed that the liquid α-benzyl derivative obtained from normal benzaldoxime contained the benzyl group attached to oxygen, and that it therefore possessed the following structure:

$$C_6H_5CH=N . O . C_7H_7$$

This naturally leads to the ordinary oxime formula for benzaldoxime. The solid β-benzyl ester, on the other hand, contained its benzyl group attached to *nitrogen;* [1] its constitution must be:

$$C_6H_5CH{\diagdown}\ \begin{matrix} N . C_7H_7 \\ | \\ O \end{matrix}$$

and that of isobenzaldoxime itself:

$$C_6H_5CH{\diagdown}\ \begin{matrix} NH \\ | \\ O \end{matrix}$$

as already determined by other reasons.

If further proof were needed, Beckmann's syntheses [2] of these esters clearly and completely prove the structures assigned to them. As is well known, hydroxylamine forms two series of substitution products — one containing radicals in place of hydroxyl hydrogen, the other in place of amido hydrogen. In

---

[1] The method employed for such determinations depends upon the almost invariable fact that alkyl groups attached to nitrogen adhere to this element with great firmness. Thus, on warming the above β-benzyl ester with hydriodic acid, benzyl-amine, $C_6H_5CH_2NH_2$, results. The α-benzyl ester, under the same circumstances, gives benzyl alcohol, $C_6H_5CH_2OH$.

[2] *Ber. d. chem. Gesell.* **22**, 1531 (1889).

the case of the benzylhydroxylamines, for example, we have the following substances :

α-benzylhydroxylamine $H_2N.O.C_7H_7$

β-benzylhydroxylamine $C_7H_7.NH.OH$

These alkylated hydroxylamines react with benzaldehyde to form the two benzyl esters; α-benzylhydroxylamine gives the liquid α-ester :

$$C_6H_5CH=\overline{O + H_2}N.O.C_7H_7 = C_6H_5CH=N.O.C_7H_7 + H_2O$$

β-benzylhydroxylamine yields the solid β-ester :

$$C_6H_5CH=O + \begin{matrix} HN.C_7H_7 \\ | \\ HO \end{matrix} = C_6H_5CH\begin{matrix} N.C_7H_7 \\ | \\ O \end{matrix} + H_2O$$

What conclusion more natural than that the oximes themselves are similarly constituted?

The Meyer-Auwers theory of the benzil dioximes was but strengthened by this demonstration of structure-isomerism on the part of the benzaldoximes; and its authors did not fail to emphasize this fact.[1] Aldoximes and ketoximes would thus appear to belong in different categories. But Beckmann's further work began to throw doubt upon the subject. It will be remembered that Auwers and V. Meyer argued the identity of the benziloximes from the identity of their oxidation products. By the same token, the benzaldoximes ought to yield *different* oxidation products, since they do not possess identical structure. However, Beckmann found that both aldoximes behave alike on being subjected to various

---

[1] *Ber. d. chem. Gesell.* **22**, 565 (1889).

oxidizing agents; [1] and he is led to the conjecture that aldoximes and ketoximes are similarly constituted.

Yet if this suggestion be true, then the two *mon*-oximes of benzil ought to behave as do the benzaldoximes. According to Auwers and Dittrich, however, this is not the case.[2]   On benzylating the former, *two oxygen*-esters are produced. Furthermore, the results of synthesis differ from Beckmann's. α-benzylhydroxylamine gives α-O-benzyl-benzaldoxime; β-benzylhydroxylamine does *not* give the γ-O-ester, but a third, *isomeric* N-ester different from either of the known ones. It is clear that these results accord better with the views of V. Meyer and Auwers than with those of Beckmann.

If the history of organic chemistry had suddenly closed at about the middle of 1889 the chronicler would have summed up the case of the oximes somewhat in this fashion : Two varieties of isomerism have been proven to exist among these substances. Aldoximes are extant in two structural modifications, distinguished by the presence of hydroxyl and imido groups.   Oximes of monoketones do not form isomeres.   Mono- and di-oximes of diketones exist in three stereomeric forms each.   The best-known case is that of benzil, of which three di- and two monoximes have been prepared.   This kind of isomerism demonstrates that carbon atoms do not rotate freely around their common axis.

But in less than six months from the period of our fictitious historian each and all of these statements

---

[1] *Ber. d. chem. Gesell.* **22**, 1588 (1889).
[2] *l.c.* **22**, 1996 (1889).

had been challenged and refuted. In the first place, H. Goldschmidt asserted the structural identity of *all* oximes, whether derived from ketones or aldehydes.[1] His argument was based upon the reactions of these substances with phenyl isocyanate. This reagent has a great affinity for hydroxyl and imido groups, with which it forms additive compounds. Thus, with alcohol it forms the corresponding phenyl-urethane :[2]

$$C_6H_5 . N{=}C{=}O + C_2H_5OH = C_6H_5 . NH . CO . OC_2H_5$$

With imido-derivatives it produces substituted ureas:

$$C_6H_5 . N{=}C{=}O + NH(C_2H_5)_2 = C_6H_5 . NH . CO . N(C_2H_5)_2$$

e.g. with diethylamine diethyl-phenylurea. Goldschmidt, to whom the extended use of this substance as a hydroxyl reagent is due, found that the oximes of benzil react normally to yield urethanes; and these urethanes are readily decomposable into their constituents, a characteristic of this class of substances. This was to be expected, at all events; but the behavior of isobenzaldoxime was perfectly similar — which was decidedly unexpected. Goldschmidt did not hesitate to declare his conviction that isobenzaldoxime possessed the same structure as normal benzaldoxime. The cause of isomerism is left undetermined. "Beckmann has proved only that the benzyl esters are structure-isomeric. From this we can draw no conclusions as to the oximes

---

[1] *Ber. d. chem. Gesell.* **22**, 3109 (1889).
[2] The esters of carbamic acid

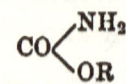

are called urethanes.

themselves.   Whatever the real cause may be, we have no basis for assuming different reasons for the isomerism in the two groups." In support of this contention Goldschmidt adduces a number of instances in which one and the same mother-substance gives two series of alkyl derivatives. (Cf. p. 72.)

The world was not obliged to wait long for an explanation of these perplexing phenomena. Two months later Hantzsch and Werner, in a purely critical and constructive treatise, solved the problem satisfactorily.[1] If we compare carbon compounds with those of nitrogen, we are struck by a certain formal analogy in the unsaturated series. Thus acetylene bears a marked resemblance to hydrocyanic acid ; we need only replace one CH group (which is of course trivalent) by a nitrogen atom :

$$\begin{array}{cc} CH & CH \\ ||| & ||| \\ CH & N \end{array}$$

Similarly, substances like maleïc acid resemble the oximes :

$$\begin{array}{cc} CX_2 & CX_2 \\ || & || \\ CX_2 & NX_2 \end{array}$$

the bivalent group $\overset{||}{CX_2}$ is replaced by the equivalent $\overset{||}{NX}$. The novel and fundamental idea introduced by Hantzsch and Werner is that in certain instances, notably the oximes, the valences of nitrogen *do not lie in a plane.* Our chief reason for accepting the doctrine of a tetrahedral carbon atom is the fact that the tetrahedron offers the simplest possible explanation of the equal spatial distribution of the four

[1] *Ber. d. chem. Gesell.* **23**, 11 (1890).

valences. But in the case of trivalent nitrogen, the simplest assumption is that the valences lie in a plane with the centre of gravity of the atom:

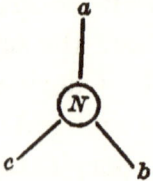

Isomeric compounds such as *Nabc* and *Nbca* would not exist if this assumption is correct. Hantzsch and Werner ascribe a tetrahedral shape (not necessarily regular) to the nitrogen atom, and thereby arrive at the same conclusions concerning the oximes as did Wislicenus about maleïc and fumaric acids (p. 163). The following diagrams will make this clear. If we regard the field of activity of a nitrogen atom as a sphere, then on the basis of plane distribution the three valences lie in the equator.:

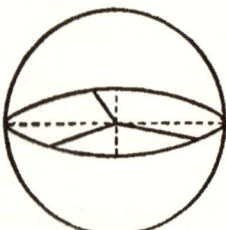

The Hantzsch-Werner conception locates the points of affinity in a circle parallel to the equator:

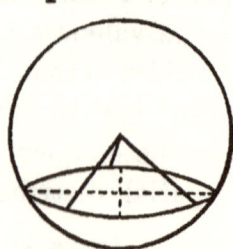

On applying this idea to the case of the oximes we obtain, *e.g.*, for the benzaldoximes, the configurations:

$$C_6H_5-C-H$$

or abbreviated:

$$C_6H_5-C-H \qquad\qquad C_6H_5-C-H$$
$$N-OH \qquad\qquad\qquad HO-N$$

The consequences entailed by this theory differ materially from those of the Meyer-Auwers hypothesis. According to Hantzsch-Werner, *all* oximes of the general formula

$$X-C-Y$$
$$NOH$$

exist in two modifications. Dioximes are to be expected in three forms, *e.g.* the benzil dioximes:

$$C_6H_5C-C-C_6H_5 \qquad C_6H_5C \underline{\qquad\qquad} C-C_6H_5 \qquad C_6H_5C\underline{\qquad}C-C_6H_5$$
$$HON \; NOH \qquad\qquad\quad NOH \; HON \qquad\qquad\quad NOH \; NOH$$

The only known case of isomeric monoximes of the above general type at the time were the benzaldoximes; and as structural isomerism was by no means absolutely precluded, V. Meyer did not fail to emphasize[1] this discrepancy of the new theory. It was perfectly clear that a well-authenticated case of this sort would deal a death-blow to the older theory. The irony of fate willed it that V. Meyer and Auwers themselves should make the first discovery. They had previously found but one oxime

---

[1] *Ber. d. chem. Gesell.* **23**, 597 (1890).

of benzophenone, a fact in accord with either view; but now it appeared that the oxime of p-chloroben-zophenone exists in two varieties.[1] The reign of the old theory is over; Hantzsch and Werner have triumphed.

The history of the oximes is thus practically closed. Meyer and Auwers[2] offered a modification of the Hantzsch-Werner hypothesis, but after a brief and unimportant discussion have accepted the rival doc-trine reservedly. Criticisms from other sides have appeared, notably Bischoff[3] and Claus[4]; but these authors have added nothing to our knowledge of the subject, and we may pass them by with this reference.

All of the consequences of the new theory have been experimentally verified, and it has even been possible to determine the spatial distribution of the atoms in these compounds. One of the most inter-esting cases is the oxime of p-tolyl-phenyl-ketone:[5]

$$(p) CH_3 . C_6H_4 . CO . C_6H_5$$

This ketone yields two oximes, and also two stereo-meric O-benzyl esters, which pass into each other as readily as do the oximes. Similar results were obtained by Goldschmidt[6] with anisaldoxime and m-nitrobenzaldoxime, each of which gives structure-

---

[1] *Ber. d. chem. Gesell.* **23**, 2403.

[2] *l.c.*

[3] *l.c.* **23**, 1970.

[4] *J. prakt. Chem.* (2), **44**, 313; **45**, 1, 377 (1891). Cf. Hantzsch, *Ber. d. chem. Gesell.* **25**, 1692 (1892).

[5] *Ber. d. chem. Gesell.* **23**, 2325, 2776 (1890).

[6] *l.c.* **23**, 2163 (1890).

identical, stereomeric O-methyl esters.   The stereo-isomerism of oxime *derivatives* is the best possible proof of the validity of a spatial hypothesis.

A word as to the nomenclature[1] of the isomeric oximes.   Hantzsch distinguishes them by the prefixes " Syn " and " Anti "; the choice is determined by the following rules: The prefix indicates the proximity of the radical attached to *nitrogen* (OH in this case) to that one of the other radicals which follows immediately after the prefix.   *E.g.*:

$$C_6H_5—C—C_6H_4 . CH_3$$
$$\|$$
$$HO—N$$
Syn-phenyltolylketoxime
or
Anti-tolylphenylketoxime

$$C_6H_5—C—C_6H_4 . CH_3$$
$$\|$$
$$N—OH$$
Anti-phenyltolylketoxime
or
Syn-tolylphenylketoxime

If one of the other radicals is capable of reacting with the group attached to nitrogen, — as *e.g.* in the aldoximes hydroxyl and hydrogen are split off in the shape of water, — the prefix is so chosen that the syllable "syn " distinguishes that one of the isomeres containing said reacting groups in closer proximity:

$$C_6H_5—C—H$$
$$\|$$
$$N—OH$$
Benzsynaldoxime

$$C_6H_5—C—H$$
$$\|$$
$$HO—N$$
Benzantialdoxime

The nomenclature of the dioximes is similar.   In the formulæ on page 184, the one with hydroxyls adjacent is called " syn," the one with hydroxyls in juxtaposition " anti," the third " amphi."

The configurations of individual oximes have been ascertained by Hantzsch according to the following

---

[1] *Ber. d. chem. Gesell.* **24,** 3479 (1891).

principles. Aldoximes, as already mentioned, easily
pass into nitriles by loss of water. That one of the
two isomeres which does so most easily evidently
contains its hydroxyl adjacent to the movable hydrogen atom; β- or isobenzaldoxime is more readily
dehydrated than the ordinary oxime, therefore it is
benz*syn*aldoxime. This method is applicable only
to aldoximes. With ketoximes, Hantzsch makes
ingenious use of the Beckmann reaction, whereby
oximes are transformed into acid amides. In the
classical case[1] of the tolyphenylketoximes, one is
converted into the anilide of toluic acid; we conclude that it contains the *phenyl* and hydroxyl radicals in the " syn " position, as the following series
will make clear:

$$CH_3.C_6H_4-C-C_6H_5 \longrightarrow CH_3.C_6H_4-C-OH \longrightarrow CH_3.C_6H_4-CO$$
$$\underset{N-OH}{\|} \qquad \underset{N-C_6H_5}{\|} \qquad \underset{NH-C_6H_5}{|}$$

The other modification is converted into the toluide
of benzoic acid, and therefore contains hydroxyl and
tolyl in the " syn " position:

$$CH_3.C_6H_4-C-C_6H_5 \longrightarrow HO-C-C_6H_5 \longrightarrow OC-C_6H_5$$
$$\underset{HO-N}{\|} \qquad \underset{CH_3.C_6H_4-N}{\|} \qquad \underset{CH_3.C_6H_4-HN}{|}$$

An objection which has been made against the
Hantzsch-Werner theory is that *all* oximes do not
form stereo-isomeres. As a matter of fact, until
within less than a year ago,[2] no stereomeric oximes
were known containing only *one* aromatic radical.

[1] *Ber. d. chem. Gesell.* **24**, 13 (1891).
[2] Scharvin, *Ber. d. chem. Gesell.* **30**, 2862 (1897).

What the cause of this curious phenomenon may be, can only be conjectured. When such "mixed" oximes undergo the Beckmann reaction, it is invariably the *aromatic* radical which changes place with the hydroxyl. We may conclude that these substances possess the configuration which this reaction implies ; *e.g.* acetophenone oxime :

$$C_6H_5-C-CH_3$$
$$\|$$
$$HO-N$$

and that for unknown reasons the other modification is too unstable for isolation by the laboratory methods with which we are acquainted. Similar circumstances obtain among the purely aliphatic oximes — no well-authenticated cases of stereoisomerism are known. A possible exception may lie in the discovery by Dunstan and Dymond[1] of two different acetaldoximes ; there is no good reason for doubting space-isomerism in this case, but the evidence is not as clear as might be wished.[2]

Hantzsch has made strenuous efforts to ascertain the conditioning factors of isomerism, but without much substantial success. He has given us a table of radicals in their order of preference for the hydroxyl group ("syn"-forming).[3] But the conditions under which the oximes must be handled in

---

[1] *J. Chem. Soc.* 1894, 206.

[2] On the other hand, v. Miller and Plöchl have come to the conclusion that the aliphatic oximes do not contain an asymmetric nitrogen atom. This is argued from the fact that these oximes add hydrocyanic acid, whereas aromatic oximes are indifferent to this reagent. Cf. *Ber. d. chem. Gesell.* 25, 2020 (1892).

[3] *Ber. d. chem. Gesell.* 25, 2104 (1892).

the laboratory (in the form of salts, esters, etc.) exercise a marked influence upon configuration, so that nothing definite can be predicted of their behavior under given circumstances.

The Hantzsch-Werner hypothesis has scored its greatest successes in the oxime group. There is nothing in the general formula:

$$\begin{array}{c} X-C-Y \\ \| \\ N-Z \end{array}$$

which should make exceptions of all other unsaturated nitrogen compounds. But strangely enough, while numerous attempts have been made to include other classes of substances under the stereochemistry of nitrogen, none of these has advanced beyond a purely tentative stage. Various isomeric hydrazones,[1] aniles[2] (condensation products of aldehydes with aniline and other aromatic bases), osazones,[3] etc.,[4] have appeared on the scene, but it has not been easy to incorporate them into the stereochemical fold with perfect consistence. As this is a branch of the subject still in its infancy, it will be best here to avoid the presentation of uncertain facts ; references to the literature in question must suffice.

One branch of the stereochemistry of nitrogen, however, has attained to sufficient proportions to warrant separate mention. During the last few

---

[1] *Ber. d. chem. Gesell.* **26**, 18 (1893); *Compt. rend.* **116**, 718 (1893); *Amer. Chem. J.* **16**, 102 (1895).

[2] *Monatash.* **14**, 279 (1893); *Ber. d. chem. Gesell.* **27**, 1297 (1894).

[3] *Gazz. chim. ital.* **26**, 444 (1896).

[4] *Ber. d. chem. Gesell.* **25**, 3098 (1892); **26**, 926, 3064 (1893).

years, a most curious and perplexing isomerism has come to light among the diazo compounds. Hantzsch has attempted to explain these phenomena on a spatial basis; with what degree of success will be seen from the detailed account given in the chapter on the constitution of diazo compounds (pp. 215 ff.).

It is a far cry from the atomic theory of Dalton to the conceptions of van t'Hoff-Le Bel and Hantzsch-Werner. Well might chemistry rest upon her laurels and point with pride to the achievements of the ebbing century. But *per aspera ad astra* has been the watchword of the organic chemist, and we may feel certain that each new generation will have its own triumphs and its own difficulties.

CHAPTER IX

THE CONSTITUTION OF THE DIAZO COMPOUNDS

In the earlier chapters, we have seen how steadily the chemistry of carbon has progressed; how, step by step, the doctrine of linked atoms has unravelled the structure of complicated substances; has expanded into the theory of ring-compounds; and, finally, was even capable of extension in a direction hardly dreamed of by its founders — witness the marvellous development of stereochemistry during the last fifteen years. The dogma of the quadrivalence of carbon has been able to withstand all attacks made upon it; all its logical consequences have been experimentally verified. It matters not that we are still unable accurately to account for the full number of valences in benzene, nor that the existence of a number of substances containing bivalent carbon has been demonstrated, — to what new heights these divergences may lead we do not know; but one fact is plain: the theory of valence, and the additional theory of a comparatively constant valence, has fully justified its existence. What wonder, then, that these theories have attempted to sway the destinies of other elements, and claimed to be the one true faith.

Indeed, at present chemists have no other hope of salvation. Without the valence theory, our speculations drift hopelessly from one quagmire to another, devoid of any certain basis. And yet, at this very moment, the chemist is tunnelling beneath the rock which alone sustains him — tunnelling among the mysteries of the compounds of nitrogen. During the last ten years, this element has been engrossing the attention of investigators in a constantly increasing degree, though each succeeding year but serves to deepen the mystery. It has long been generally understood that, in its compounds, nitrogen is either trivalent or pentavalent; but this tacit assumption begins to show signs of distress, and the very idea of valence itself is threatened.[1] The battle rages all along the line of nitrogen chemistry, but chiefly among the diazo compounds, the first to draw attention to the vagaries of the nitrogen atom.

The diazo compounds were discovered in 1858 by Peter Griess.[2] The first detailed description of them was published two years later,[3] and the following accounts were scattered throughout the next six years; though occasional notes from Griess's pen occur for twenty years more. It will be best in this place to recount Griess's achievements descriptively rather than historically, even though thereby we lose sight of the wonderful beauty of his experimental investigations. It is difficult to convey the attitude of contemporary workers toward these masterly researches. Not alone did the absolute novelty of the substances described by Griess arouse amazement, but the consummate skill, the

---

[1] Cf. Brühl, *Ber. d. chem. Gesell.* **31**, 1350 (1898).

[2] *Ann. Chem.* (Liebig), **106**, 123 (1858).

[3] *l.c.* **113**, 201 (1860).

mastery of laboratory technique, the versatility displayed during the progress of the investigation, called forth the enthusiastic admiration of every competent critic, — a homage which the recent developments in the field of diazo chemistry have served but amply to confirm.

As is well known, when nitrous acid acts upon amino compounds, the amino group is replaced by hydroxyl:

$$R-NH_2 + HNO_2 = R-OH + N_2 + H_2O$$

This reaction was discovered by Piria,[1] who by this means converted asparagine into malic acid. Certain observations made in Kolbe's laboratory led that brilliant chemist to suspect the existence of an intermediate compound in Piria's reaction when applied to aromatic amines, and, at his suggestion, Griess undertook the investigation. Griess isolated his first *diazo compound*, as he decided to call the new class of substances, by working with alcohol as a solvent; subsequently he was able to dispense with this experimental aid, for the early experiences with his new substances taught him how they might be obtained under simpler circumstances. In short, Griess found that whenever nitrous acid acts upon aromatic amines, no matter what the solvent nor how the nitrous acid is introduced, a diazo compound is formed, provided the temperature be kept below the decomposition-point of the new substance. His first and favorite means of employing nitrous acid in these reactions was to pass the gas, commonly known as nitrogen trioxide, into the given solutions;

[1] *Ann. Chem.* (Liebig), **68**, 348 (1846).

but any soluble nitrite could be substituted, the acid being liberated by a stronger mineral acid ; at times, Griess even used ethyl and amyl nitrites for this purpose. If the solid reaction-product was required, concentrated solutions were taken, or by addition of suitable solvents it was precipitated in crystalline form ;[1] but fortunately, for most purposes, a dilute aqueous solution is sufficient or even preferable.

The diazo compounds are characterized by extreme instability and by the variety of their reactions. In the solid state most of them are explosive, diazobenzene nitrate being the most dangerous of all. In solution they quickly decompose, giving off nitrogen, and forming complicated products of unknown composition. Under definite circumstances, however, the decomposition may be made to take definite directions.

The analysis of these substances proved a matter of no small difficulty. The first few diazo compounds happened to be comparatively stable, so that their composition could be determined fairly safely; with the information thus gained, Griess was able to make skilful use of a decomposition reaction to determine the composition of the other members of the family. Thus, when diazobenzene nitrate is warmed with dilute sulphuric acid, it breaks down into nitric acid, phenol, and nitrogen :

$$C_6H_5N_2NO_3 + H_2O = HNO_3 + N_2 + C_6H_5OH$$

---

[1] A better method has been given by Knoevenagel (*Ber. d. chem. Gesell.* 23, 2994). It consists in employing alcohol as a solvent, and amyl nitrite as the diazotizer.

By determining the amount of acid set free, and above all the quantity of nitrogen given off, it is possible to accurately gauge the composition of the substance. The estimation of diazo nitrogen thus becomes one of the most important analytical methods in this branch of chemistry.

The reactions of the diazo compounds are extremely varied. As might be expected of a substance containing so much nitrogen, the diazo complex is strongly base-forming. Griess prepared numberless salts not only of diazobenzene itself, but of a great quantity of substituted diazobenzenes, diazotoluenes, etc.[1] Examples are diazobenzene nitrate, chloride, sulphate :

$$C_6H_5N_2.NO_3 \qquad C_6H_5N_2.Cl \qquad (C_6H_5N_2)_2SO_4$$

diazo-p-toluene nitrate :

$$C_6H_4\begin{cases} CH_3 \\ N_2.NO_3 \text{ (p)} \end{cases}$$

p-brom-diazobenzene bromide : .

$$C_6H_4\begin{cases} N_2.Br \\ Br \text{ (p)} \end{cases}$$

As a rule, in preparing a diazo compound such a salt is obtained first, which then serves as the starting-point for the numerous reactions which characterize this class of substances. The general feature of most of these reactions is the elimination of nitrogen from the molecule. Thus, all of the salts mentioned above, when boiled with a dilute acid (preferably

[1] *Ann. Chem.* (Liebig), 137, 39 (1866).

dilute sulphuric acid), lose all their diazo nitrogen, and form phenols:

$$C_6H_5N_2 . Cl + H_2O = C_6H_5OH + N_2 + HCl$$

This simply carries to its conclusion Piria's reaction, whereby the action of nitrous acid upon amines produces alcohols. If in the above decomposition dilute acids be replaced by concentrated ones, the reaction takes an entirely different course. Concentrated hydrochloric acid yields chlorbenzene:

$$C_6H_5N_2 . NO_3 + HCl = C_6H_5Cl + N_2 + HNO_3$$

hydrobromic and hydriodic acids act similarly:

$$C_6H_5N_2 . Cl + \begin{cases} HBr = C_6H_5Br + N_2 + HCl \\ HI = C_6H_5I + N_2 + HCl \end{cases}$$

By this means we are able to replace any amido-group in an aromatic molecule by one of the halogens; and since nitro- (and therefore amido-) groups are much more easily introduced into the benzene ring than chlorine or bromine, these diazo reactions of Griess are valuable laboratory adjuncts.[1]

But the decomposition of diazo compounds by means of acids does not exhaust the capabilities of this Protean class of substances. If they are boiled with alcohol (preferably in the presence of an alkali), the diazo complex is eliminated entirely and replaced

---

[1] The value of this series of reactions has been greatly augmented by a discovery of Sandmeyer (*Ber. d. chem. Gesell.* 17, 1633 (1884)). In the presence of cuprous salts, the replacement of the diazo group takes place much more readily; we may even introduce the cyanogen group in this way, and can thus build up aromatic acids. For details of this method, and of a modification by Gattermann, consult any larger text-book or laboratory manual.

by hydrogen ;[1] the alcohol becomes oxidized to aldehyde during this process :

$$C_6H_5N_2 . NO_3 + C_2H_6O = C_6H_6 + N_2 + HNO_3 + C_2H_4O$$

A by-product of the reaction is the hydrocarbon diphenyl, evidently formed by the union of two nascent phenyl groups :

$$2 C_6H_5N_2 . NO_3 + C_2H_6O = (C_6H_5)_2 + 2 N_2 + 2 HNO_3 + C_2H_4O$$

Even this elimination reaction has its value; *e.g.* in the preparation of meta-toluidine.[2]   The direct nitration of toluene yields a mixture of the ortho- and para-nitrotoluenes, which upon reduction of course give the corresponding toluidines.   The meta-derivative cannot be prepared in this way.   But if we nitrate para-toluidine, the entering nitro group takes up the meta-position with reference to the methyl group :

The amido group is then eliminated as just described, and there remains meta-nitro-toluene.

A further curious reaction is given by diazobenzene when treated with bromine : a compound known as diazobenzene perbromide is formed :

$$C_6H_5N_2Br_3$$

---

[1] In the absence of alkali, the reaction takes another course :

$$C_6H_5N_2 . Cl + C_2H_5OH = C_6H_5OC_2H_5 + HCl + N_2$$

Phenol ethers are thus formed.

[2] A full account of this process is given in Meyer-Jacobson, II, 155.

This, when warmed, breaks down into nitrogen, bromine, and brombenzene:

$$C_6H_5N_2Br_3 = C_6H_5Br + N_2 + Br_2$$

Treated with ammonia it yields another strange substance: the three bromine atoms are replaced by one atom of nitrogen:

$$C_6H_5N_2 \overline{| Br_3 + H_3 |} N = 3\,HBr + C_6H_5N_3\,;$$

the substance is known as diazobenzene imide, and forms a volatile, explosive oil. Diazobenzene imide, when reduced by zinc and sulphuric acid, finally breaks down into aniline and ammonia:

$$C_6H_5N_3 + 8\,H = C_6H_5NH_2 + 2\,NH_3$$

We have not yet reached the end of Griess's work with the diazo compounds; but we must pause for a moment to consider the constitution of these substances from a contemporary standpoint. Griess himself did not ponder much over the problematical side of the subject; in his first paper he tells us all he has to say about it, though he subsequently adhered to his theory with stubborn tenacity. To fully comprehend his point of view, we must remember that in 1860 there was no "benzene theory" upon which to base other hypotheses. Looking, then, at the formula of, *e.g.*, diazobenzene nitrate,

$$C_6H_5N_3O_3$$

Griess reasoned as follows: all ammonia bases, such as aniline, methylamine, ammonia itself, form salts with acids by direct addition; the formula of diazobenzene nitrate therefore can be resolved as follows :

$$C_6H_4N_2 . HNO_8$$

The complex $C_6H_4N_2$ is the group or radical which corresponds to ammonia or aniline; it may be regarded as aniline in which three hydrogen atoms are replaced by one atom of nitrogen, or as benzene in which two hydrogen atoms are replaced by two atoms of nitrogen. The latter view was the one first adopted by Griess; subsequently he exchanged it for the former. Both forms of the theory involved, however, that the group N_2 adhered to the benzene ring in *two* places; but Griess never felt himself able to specify the second place. These views were probably adopted without discussion until the birth of the benzene theory of Kekulé, when that brilliant speculator advanced a conception of the diazo compounds which completely overthrew Griess's vague, unsatisfactory assumptions.

Kekulé based his formula[1] upon his views of valence and upon his new benzene theory. Nitrogen to him was trivalent; therefore two nitrogen atoms could not well replace two atoms of hydrogen. Moreover, the benzene radical is always C_6H_5 for monosubstituted derivatives. Diazobenzene nitrate therefore must be formulated as follows :

$$C_6H_5 . N=N . NO_8$$

This constitution explains why the decomposition reactions of diazobenzene salts always yield monosubstitution products of benzene; if, according to

[1] *Lehrbuch*, II, 703 (1866).

Griess, the diazo complex is attached in two places, substitution ought to occur twice. Moreover, we get a simple, comprehensible formula for diazobenzene perbromide :

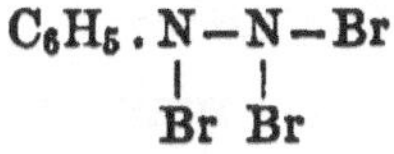

$$C_6H_5 . N-N-Br$$
$$\quad\quad\; |\;\;\; |$$
$$\quad\quad Br\;\; Br$$

and for diazobenzene imide :

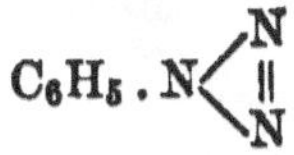

$$C_6H_5 . N \underset{N}{\overset{N}{<}} ||$$

Kekulé's views are distinctly superior to Griess's, and were universally adopted — for the time being, at any rate — except by Griess himself.

We can now return to our contemplation of Griess's experimental investigations. An unexpected turn was given to the chemistry of nitrogen by the discovery that diazobenzene has *acid* properties. Griess is very much struck by the fact that while the introduction of a second nitrogen atom invests the molecule with decidedly greater basic properties (diazobenzene is a much stronger base than aniline), the complex is also capable of uniting with metals to form substances resembling salts. Thus, when diazobenzene nitrate is treated with strong caustic potash,[1] the following reaction takes place :

$$C_6H_5N_2NO_3 + 2\,KOH = C_6H_5N_2OK + KNO_3 + H_2O$$

the products being potassium nitrate and diazobenzene potassium. Solutions of this new potassium

---

[1] *Ann. Chem.* (Liebig), **137**, 53 (1866).

salt, treated with silver nitrate, give a precipitate of diazobenzene silver:

$$C_6H_5N_2OK + AgNO_3 = C_6H_5N_2OAg + KNO_3$$

Griess's analyses of these compounds were branded as questionable by Curtius[1] many years later. Curtius was unable to prepare substances having the composition given by Griess. However, Schraube and Schmidt[2] have recently shown that a slight modification of Curtius's method will indeed yield pure metal salts of diazobenzene; and thus we may feel quite certain that Griess made sure of his facts before publishing them. Diazobenzene potassium, when treated with dilute acetic acid, yields a very unstable oil, which Griess looked upon as free diazobenzene; he formulated it thus:

$$C_6H_4N_2$$

whereas Kekulé gave it the following constitution:

$$C_6H_5N_2 . OH$$

As a matter of fact, the oil was not diazobenzene at all, as Curtius's[3] subsequent inquiry proved; its nature is still a mystery.

To sum up the differences between Griess's and Kekulé's diazobenzene formulæ once more, the following brief table will serve to show the chief distinctions in the two views:

[1] *Ber. d. chem. Gesell.* **23**, 3035 (1890).

[2] *l.c.* **27**, 520 (1894).

[3] *l.c.*

Griess	Kekulé	
$C_6H_4N_2$	$C_6H_5N_2 . OH$	Free diazobenzene
$C_6H_4N_2 . HCl$	$C_6H_5N_2 . Cl$	Diazobenzene chloride
$C_6H_4N_2 . KOH$	$C_6H_5N_2 . OK$	Diazobenzene potassium
$C_6H_4N_2 . HBr . Br_2$	$C_6H_5NBr . NBr . Br$	Diazobenzene perbromide[1]
$C_6H_4N_2 . NH$	$C_6H_5N{-}{-}N$	Diazobenzene imide

Griess's formulæ have made no headway against Kekulé's, and we need not concern ourselves with them further. The main difference lies in Kekulé's conception of diazobenzene as a base corresponding to ammonium hydroxide rather than to ammonia, as the following equation shows:

$$C_6H_5N_2 . OH + HNO_3 = C_6H_5N_2 . NO_3 + H_2O$$

But we are not yet done with our study of Griess's discoveries. Among the first of his diazo compounds were the diazophenols, formed by diazotizing amidophenols. These diazo derivatives are *neutral* substances, a fact very well expressed by Kekulé's formula of intramolecular anhydrides (analogous to betaine[2]):

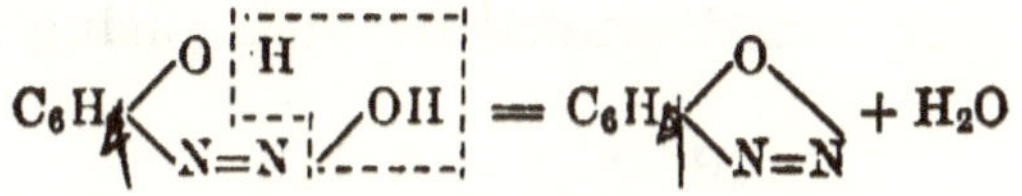

[1] Very recent investigations by Hantzsch have shown that this substance is a genuine perbromide, and that its constitution is probably

$$C_6H_5N_2Br . Br_2,$$

corresponding to cæsium perbromide, $CsBr . Br_2$.

[2] Betaïne is an intromolecular anhydride of trimethyl hydroxyammoniumacetic acid:

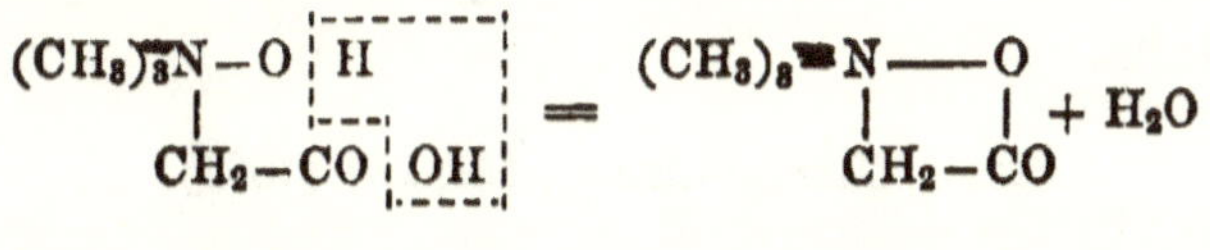

Another class of substances was brought to light when Griess allowed *two* molecules of an amine to act upon one of nitrous acid:

$$2 C_6H_5NH_2 + HNO_2 = (C_6H_5)_2N_3H + 2 H_2O$$

This substance was christened *diazo-amido-benzene;* its investigation was a particularly brilliant achievement, for the regular methods of analysis were rendered impossible by the explosiveness of the substance. Griess found,[1] however, that when treated with concentrated hydrochloric acid, diazo-amidobenzene yielded equal molecular quantities of aniline and of chlorbenzene:

$$(C_6H_5)_2N_3H + HCl = C_6H_5Cl + N_2 + C_6H_5NH_2$$

that the amount of nitrogen given off on warming with dilute acids corresponds to a molecular combination of diazobenzene and aniline; and that finally diazoamidobenzene is produced when aniline and diazobenzene salts are brought together. The structure of the substance is therefore expressed by the following constitutional formula (using Kekulé's version):

$$C_6H_5 . N=N-NH . C_6H_5$$

The decomposition with acids is shown thus:

$$\overset{Cl \,|\, H}{C_6H_5 . N=N-NH . C_6H_5}$$

the molecule breaking down into aniline and diazobenzene, the latter then undergoing its own peculiar reactions.

[1] *Ann. Chem.* (Liebig), **121**, 257 (1862).

It is difficult to find a brighter bit of close chemical reasoning in the whole field of structural chemistry than this elucidation of the structure of diazoamido-benzene by Griess at a time when the resources of the organic chemist were decidedly limited. Here again we find Griess carefully following up all the consequences of a given assumption. Diazoamido compounds being formed by the union of a diazo compound with a base, it should make some difference which base and which diazo compound are employed to synthesize a given diazoamido deriva-tive. Thus, the substance formed from aniline and bromdiazobenzene :

$$C_6H_5-NH_2 + Cl\ N=NC_6H_4Br = C_6H_5-NH-N=N-C_6H_4Br + HCl$$

ought to differ from the diazoamido compound formed by union of diazobenzene and bromaniline :

$$C_6H_4BrNH_2 + Cl\ N=NC_6H_5 = C_6H_4Br-NH-N=N-C_6H_5 + HCl$$

The difference in structure consists in the different location of a hydrogen atom. As a matter of fact, Griess[1] found the two substances to be *identical*. This was one of the first instances of a phenomenon now quite familiar to us, viz. tautomerism. Laar's well-known hypothesis (cf. p. 86) arose from an attempt to account for the peculiarities of diazoamido compounds. It may be said in passing that the sub-ject of tautomeric diazoamido and other closely allied substances is engrossing considerable attention at present.

[1] *Ber. d. chem. Gesell.* **7,** 1618 (1874).

But the end of the diazo reactions is not yet. When diazobenzene solutions are brought into contact with phenols or tertiary aromatic amines, derivatives of azobenzene[1] are formed; in the one case oxy-azo-compounds:

$$C_6H_5N_2Cl + C_6H_5OH = C_6H_5-N=N-C_6H_4OH + HCl$$

in the other amido-azo-compounds:

$$C_6H_5N_2Cl + C_6H_5N(CH_3)_2 = C_6H_5-N=N-C_6H_4N(CH_3)_2 + HCl$$

These two derivatives are mother-substances to an enormous number of the so-called azo-dyes. In this connection we may refer to a curious method of forming amidoazobenzene derivatives from diazo-amido compounds. Kekulé[2] found that diazoamido-benzene, when warmed with a small quantity of an aniline salt, is completely rearranged into amidoazo-benzene:

$$C_6H_5-N=N-NH-C_6H_5 \longrightarrow C_6H_5-N=N-C_6H_4-NH_2$$

This is a perfectly general reaction for all diazoamido compounds. It resembles a number of similar cases; *e.g.* the formation of benzidine from hydrazobenzene:

$$C_6H_5-NH-NH-C_6H_5 = H_2N-C_6H_4-C_6H_4-NH_2$$

the common feature of nearly all such intramolecular rearrangements being the formation of para-substitu-

---

[1] The nomenclature of azo and diazo compounds is a trifle confusing. *Azo* compounds contain the group $-N=N-$, linked to *two* carbon atoms; *e.g.* azobenzene:

$$C_6H_5 - N = N - C_6H_5.$$

Diazo compounds, on the other hand, contain the group $C - N = N - X$, where X represents any element except carbon.

[2] *Ztschr. Chem.* 1866, 689.

tion products.   No satisfactory explanation of these remarkable molecular tornadoes has been advanced.

We owe one more curious class of substances to Griess.   When an ortho-diamido-benzene derivative is diazotized, *e.g.* o-phenylene diamine,[1] there is formed a sort of intramolecular diazoamido compound; one amido group becomes diazotized, and immediately hitches up with the neighboring amido group :

$$C_6H_4 \begin{cases} NH_2(o) \\ N{=}N.OH \end{cases} = C_6H_4 \begin{cases} NH \\ N{\diagdown}N \end{cases} + H_2O$$

Griess assigns the following constitution to these *azimido* compounds:

$$C_6H_4 \begin{cases} N \\ | \\ N \end{cases} NH$$

which he defended against Zincke's[2] advocacy of

$$C_6H_4 \begin{cases} N \\ NH \end{cases} N$$

For a long time Griess seemed to have the advantage, but of late it would appear that Zincke's formula is a better expression of the behavior of the substituted azimidobenzenes.

We now pass to a second stage in the history of diazobenzene.   Heretofore Kekulé's formulæ have reigned supreme, Griess alone standing aloof.   But a new structure puts forth its claims.   Blomstrand[3] feels a discrepancy between the constitutional for-

---

[1] *Ber. d. chem. Gesell.* 5, 200 (1872); 15, 1878 (1882).
[2] Cf. *Ann. Chem.* (Liebig), 291, 393 (1896).
[3] Cf. *Ber. d. chem. Gesell.* 8, 51 (1875).

mula of diazobenzene and those of other nitrogen compounds. Diazobenzene salts, he contends, are true ammonium salts. In all ammonium salts we postulate the presence of pentavalent nitrogen; witness ammonium chloride, tetra-ethylammonium iodide, etc.:

$$H_4 \equiv \overset{v}{N} - Cl \qquad (C_2H_5)_4 \equiv \overset{v}{N} - I$$

No case is known, says Blomstrand, in which trivalent nitrogen assumes such a rôle as does the nitrogen of diazobenzene. We may therefore regard diazobenzene chloride, for example as aniline hydrochlorate :

$$C_6H_5 - N \underset{Cl}{\overset{H_3}{\lessgtr}}$$

in which the three ammonia hydrogens have been replaced by an equivalent nitrogen atom :

$$C_6H_5 - N \underset{Cl}{\overset{N}{\lessgtr}}$$

All diazo compounds in which the characteristic basic reactions persist are formulated similarly by Blomstrand.

At about the same time, though without knowledge of Blomstrand's publication, Strecker[1] advanced the same formula, albeit for entirely different reasons. The atomic linking which Kekulé assumes in diazobenzene is also present in the well-known substance azobenzene :

$$C_6H_5 - N = N - C_6H_5$$

[1] *Ber. d. chem. Gesell.* **4,** 786 (1871).

It is reasonable to suppose that similarity of structure implies similarity of chemical behavior. Now, diazobenzene and its compounds are extremely unstable, whereas azobenzene and its derivatives are quite stable. Since there can be little doubt that azobenzene actually has the formula ascribed to it, Strecker advises the adoption of Blomstrand's modification as the best expression of the chemical peculiarity of diazobenzene. A few years later the same ideas were elucidated by Erlenmeyer[1] in somewhat greater detail, without any essential additions or modifications.

It cannot be said that these somewhat radical views made much impression upon the chemical world. They were barely discussed; chemists believed one way or the other, much as they felt inclined. The direct formation of azo derivatives (cf. p. 205) from diazobenzene salts did not agree well with the new theory, and various assumptions as to rearrangement during this reaction became necessary. And finally, an experimental investigation by Emil Fischer seemed to triumphantly vindicate the older formula.

Fischer[2] found that when diazobenzene solutions act upon neutral potassium sulphite, there is formed a diazobenzene sulphonate :[3]

$$C_6H_5N=NCl + K_2SO_3 = C_6H_5-N=N-SO_3K + KCl$$

If potassium acid sulphite is taken, phenylhydrazine

[1] *Ber. d. chem. Gesell.* 7, 1110 (1874).
[2] *Ann. Chem.* (Liebig), 190, 73 (1877).
[3] This is distinguished from the following one as Fischer's salt.

sulphonic acid[1] is produced :

$$C_6H_5N{=}N-Cl + 2\,KHSO_3 + H_2O = C_6H_5\text{-}NH\text{-}NH\text{-}SO_3K$$
$$+ KCl + H_2SO_4$$

These two salts stand to each other in the relation expressed by the formulæ, because the latter can be prepared by direct reduction of the former, the former by direct oxidation of the latter. Fischer's salt, when hydrolyzed by strong mineral acids, breaks down into a diazobenzene salt and potassium acid sulphite :

$$C_6H_5N{=}N \mid SO_3K + HCl = C_6H_5N{=}NCl + KHSO_3$$

Strecker's salt, on the other hand, yields potassium acid sulphate and phenylhydrazine :

$$C_6H_5NH-NH{+}SO_3K + H_2O = C_6H_5NH-NH_2 + K\,HSO_4$$

Now, reasoned Fischer, from the constitution of phenylhydrazine we may determine that of diazobenzene. According to whether we start with the Kekulé or with the Blomstrand formula, we obtain the following possible structures of phenylhydrazine :

Kekulé	Blomstrand	
$C_6H_5N{=}N-SO_3K$	$C_6H_5N-SO_3K$ $\underset{N}{\overset{|||}{}}$	Diazosulphonate
$C_6H_5NH-NH-SO_3K$	$C_6H_5NH-SO_3K$ $\underset{NH}{\overset{||}{}}$	Hydrazine-sulphonate
$C_6H_5NH-NH_2$	$C_6H_5NH_2$ $\underset{NH}{\overset{||}{}}$	Phenylhydrazine

The formula which Fischer deduces for phenylhydrazine is the one derived from Kekulé's diazobenzene

---

[1] The salt of this acid is known as Strecker's salt.

structure.   The chief argument is that the secondary hydrazines, whose formula must be

$$C_6H_5 \underset{R}{\diagup} N - NH_2$$

behave just like phenylhydrazine itself toward most reagents; they must therefore have similar structures.   Another argument depends upon the products formed by action of ethyl bromide upon phenylhydrazine.

At the time Fischer's "hydrazine" standpoint seemed convincing and final.   But it is a striking instance of the change which has come over us to realize that to-day we can attach but little importance to his proofs.   Since we have learned to estimate the nature and the value of addition reactions, we are much more careful in forming conclusions about molecular structures.   It would take a bold chemist indeed to say now just what differences in behavior are to be expected of substances related as are the two phenylhydrazine formulæ considered above.   One formula passes into the other by the simple shifting of a hydrogen atom; witness the havoc wrought with structural conceptions in the case of acetoacetic ether (cf. p. 87).   The same plea might now be made for the Blomstrand formula for diazobenzene; once we assume the intermediate formation of an addition product in the formation of azo-dyes, one formula is as good as the other.

And this might have been the final word in the diazo chapter were it not that the last five years have brought us wealth of new material, and, it must be said, a storm of error and refutation, of abuse and

counter-abuse.   The chemical world has seldom witnessed a fiercer struggle among contending theorists
than that recorded in the pages of recent diazo history.   We cannot here concern ourselves with the
details of a controversy as yet hardly ended, nor with
the many mistakes and retractions made on both
sides.   We must be content to hear each side say its
say, and leave the final decision to the future.

It is said of certain periods of scientific activity
that this or that discovery "was in the air."   This
theory of scientific contagion must certainly be applicable to the last stage of diazo chemistry, for on two
occasions important discoveries reached the light
simultaneously.   The first palpable intimation of
dissatisfaction with Kekulé's formula appeared from
von Pechmann,[1] who assumed that diazobenzene
reacts as the tautomeric phenylnitrosamine:

$$C_6H_5-N=N-OH \longrightarrow C_6H_5-NH-NO$$

Diazobenzene in alkaline solution reacts with ketones; *e.g.* acetoacetic ether, to form so-called
"mixed," *i.e.* part fatty, part aromatic, azo substances:

$$CH_3CO-\underset{\underset{N=N-C_6H_5}{|}}{CH}-COOC_2H_5$$

The investigation of these mixed azo compounds, in
which V. Meyer,[2] von Pechmann, and, above all,
Japp and Klingemann,[3] took part, brought out the

---

[1] *Ber. d. chem. Gesell.* **21**, 11 (1888).

[2] *l.c.* **25**, 3190 (1892).

[3] *Ann. Chem.* (Liebig), **247**, 190 (1888).

fact that the name was a misnomer; that in reality these substances are hydrazones,

$$CH_3CO-\underset{\underset{N-NHC_6H_5}{\|}}{C}-COOC_2H_5$$

and *not* azo bodies. The nitrosamine formula of diazobenzene, according to von Pechmann, readily explains this:

$$CH_3CO-C\,\vert\,H_2\,\vert\,-COOC_2H_5$$
$$\vert\,O\,\vert\,N-NHC_6H_5$$

Pursuing this idea experimentally, von Pechmann,[1] and, at the same time, Wohl,[2] found that diazobenzene, treated with benzoyl chloride, yields nitrosobenzanilide; a fact easily accounted for by the nitrosamine formula:

$$C_6H_5N\,\vert\,H\,\vert-NO+\vert\,Cl\,\vert\,COC_6H_5\ =\ \underset{COC_6H_5}{\overset{C_6H_5N-NO}{\vert}}+HCl$$

It is true, by assuming an addition reaction, it is possible to account for the formation of nitrosobenzanilide with the regular diazo structure, but the interpretation seems a trifle forced.

A little later Bamberger[3] began the study of the oxidation products of diazobenzene. These were nitrosobenzene, nitrobenzene, azobenzene, diphenyl, and a substance called diazobenzenic acid. This substance seems to possess the structure of phenylnitramine:

$$C_6H_5NH-NO_2$$

[1] *Ber. d. chem. Gesell.* **25**, 3199 (1892).
[2] *l.c.* 3631.   [3] *l.c.* **26**, 471 (1893).

Its formation, however, can be readily accounted for by any diazobenzene formula, albeit easiest by that of phenylnitrosamine.

The relation of diazobenzenic acid to diazobenzene assumed sudden importance, and at the same time the third era in diazo history was begun, when three different investigators simultaneously discovered *isomeres* of the well-known diazo compounds of Griess. Schraube and Schmidt[1] found that when solutions of paranitrodiazobenzene are treated with an alkali, a salt is precipitated to which they ascribe the formula :

$$(p)\ (NO_2)C_6H_4N{<}^{Na}_{NO}$$

This p-nitrophenylnitrosamine sodium, when decomposed by acetic acid, yields a substance isomeric with p-nitrodiazobenzene, and which they regard as p-nitrophenylnitrosamine itself :

$$C_6H_4{<}^{NO_2}_{NH-NO}$$

This substance is much more stable than ordinary diazo compounds usually are ; it is difficultly soluble in water, and, above all, neither it nor its sodium salt gives a reaction characteristic of all true diazo compounds, viz. to form azo-dyes with phenols. If, however, a mineral acid be added to the solution, the nitrosamine is isomerized to the diazo compound, and this will upon addition of alkali react normally to form an azo-dye.[2] This gives a means for recogniz-

---

[1] *Ber. d. chem. Gesell.* **27**, 514 (1894).

[2] A favorite substance for such tests is the so-called " R-salt," which is a β-naphtholdisulphonic acid (2 : 3 : 6).

ing "isodiazo" compounds; they will give the usual diazo reaction only after treatment with a mineral acid, which isomerizes them.

Schraube and Schmidt were not able to prepare the mother-substance of the isodiazobenzene series, isodiazobenzene itself; but they succeeded in converting Griess's diazobenzene potassium into the isomeric isodiazobenzene potassium (by means of hot, concentrated potassium hydroxide solutions), and found it to possess the characteristic isodiazo[1] reaction.

The chief reasons for assigning the nitrosamine formula to the isodiazo compounds are briefly thus: In the first place, p-nitroisodiazobenzene sodium reacts with methyl iodide to form phenylmethylnitrosamine :

$$C_6H_5N \begin{cases} CH_3 \\ NO \end{cases}$$

and secondly, the fact that the nitrosamine formula had recently manifested itself as the "tautomeric" structure of diazobenzene. However, von Pechmann and Frobenius,[2] who had also discovered the isomerization of p-nitrodiazobenzene, found that the silver salt reacted differently with methyl iodide, producing not a nitrosamine, but the methyl ester of p-nitrodiazobenzene :

$$C_6H_5N=N.OCH_3$$

But this merely showed that diazo compounds, like hydrocyanic acid, are true tautomeric substances.

---

[1] Even carbonic acid converts this isodiazo compound into normal diazobenzene.

[2] *Ber. d. chem. Gesell.* **27**, 672 (1894).

A third reason for the nitrosamine formula was adduced by Bamberger, who had also stumbled upon the isodiazo compounds. Bamberger found that while normal diazo compounds are with difficulty oxidized to diazobenzenic acid (which he regarded as phenylnitramine), the iso-derivatives are easily and almost completely oxidized. This is very much clearer on the basis of the nitrosamine formula for the latter, since nitroso compounds are usually easily oxidized to nitro bodies:

$$C_6H_5N\diagdown^{H}_{NO} \longrightarrow C_6H_5N\diagdown^{H}_{NO_2}$$

All of a sudden a bombshell was dropped into the diazo camp. Hantzsch,[1] in an elaborate and somewhat pretentious treatise, subjected the whole recent history of the subject to a critical review, and arrived at the following conclusions: In the first place, there is complete parallelism in the history of the diazo compounds and the oximes. In the second place, there are no fully valid reasons for the nitrosamine formula of the isodiazo compounds; the different behavior of the sodium and silver salts when alkylated simply shows tautomerism; and, moreover, when the compounds of von Pechmann and Wohl (p. 212) are saponified, Bamberger[2] obtained *normal* instead of isodiazobenzene. Thirdly, normal and isodiazo compounds have identical structure. And fourthly, since they are structurally identical, the cause of their isomerism must be stereochemical. The experimental basis of this

---

[1] *Ber. d. chem. Gesell.* 27, 1702 (1894).   [2] *l.c.* 914.

new view was slight; it consisted of nothing more than the isolation of a new diazobenzene sulphonate (of potassium), isomeric with Fischer's salt (p. 208). The new salt is very unstable, decomposes easily, and on standing soon passes into the stable, older isomere. Hantzsch took the fact for granted that these two salts were structurally identical, and deduced for them the following space formulæ :

$$C_6H_5 . N \qquad\qquad C_6H_5 . N$$
$$\| \qquad\qquad\qquad \|$$
$$SO_3K . N \qquad\qquad N . SO_3K$$
Syn-diazo compound   Anti-diazo compound

The stereochemistry of the diazo compounds is then developed by Hantzsch on the basis of that of the oximes (which see).   It will be remembered that one of the postulates of the Hantzsch-Werner hypothesis was that asymmetry being caused by a nitrogen double linking, it should be expected among such compounds as hydrazones, azo compounds, etc. :

$$C_6H_5NH . N{=}R \qquad C_6H_5N{=}NC_6H_5$$

Owing to the presence of two such nitrogen atoms in azo and diazo compounds, isomerism of this sort was all the more likely.   The discovery of isodiazo derivatives seemed the verification of the predictions of 1890.   With reference to the case in question, it need only be added that Hantzsch grouped the normal diazo compounds, with their tendency toward spontaneous decomposition, among the syn-derivatives :

$$C_6H_5N \qquad\quad C_6H_5N \qquad\quad C_6H_5N$$
$$\| \qquad\qquad\quad \| \qquad\qquad\quad \|$$
$$Cl . N \qquad\quad NO_3 . N \qquad\quad HO . N$$

and the more stable isodiazo compounds among the anti-derivatives :

$$(NO_2)C_6H_4N \qquad C_6H_4ClN$$
$$\| \qquad\qquad \|$$
$$NOH \qquad\qquad NONa$$

And he added a theorem to the effect that "syndiazo compounds *couple*,[1] antidiazo compounds do not," this being the stereochemical formulation of the above-described "isodiazo" reaction.

It cannot be denied that this publication produced a deep impression upon the chemical public, an impression which was heightened by a further communication by Hantzsch, in which he described an isomeric diazoamidobenzene.[2] Structurally speaking, an isomeric diazoamidobenzene is impossible, and a spatial conception offered the only explanation.

The first manifestation of this impression was a vigorous protest by Bamberger.[3] The details of this careful arraignment need not be given here ; suffice it to say that Bamberger showed conclusively that Hantzsch had not proved the assumed identity of structure in his and Fischer's diazosulphonate, and that the latter salt, which according to Hantzsch belongs to the iso series, does not give the isodiazo reaction. Bamberger, for his part, is convinced that the nitrosamine formula amply accounts for all known reactions of the isodiazo compounds. As to Hantzsch's isomeric diazoamidobenzene, it was easy

---

[1] Coupling (*Kuppeln*) is the term employed to designate reactions in which two molecules hitch together, as in the formation of azo-dyes.

[2] *Ber. d. chem. Gesell.* **27**, 1857 (1894).

[3] *l.c.* 2582.

to show that it did not exist,[1] and that the elaborate superstructure of conclusions he had based upon it was therefore valueless. The isomeric sulphonates are declared to be isomeric only with reference to the " inorganic " part of the molecule :

$$X-SO_3K \qquad X-O-SO_2K$$

The more recent history of the diazo compounds is too complicated for a brief summary treatment. In large part the details are uncertain, or even flatly contradict each other. In this place we can do no more than record as much as may be regarded as positive facts, and let the opposing statements and contentions get on as best they can. Judgment has not yet been rendered,[2] and until then we must be patient.

We must first notice a view of the isomeric sulphonates advanced by eminent critical authorities. V. Meyer and P. Jacobson, in their invaluable *Lehrbuch*, express the conjecture that the differences between Fischer's and Hantzsch's salts may be explained by the following formulæ :

$$C_6H_5N=N-SO_3K \qquad C_6H_5N{\displaystyle{\nearrow N \atop \searrow SO_3K}}$$

the former derived from Kekulé's diazo formula, the latter from Blomstrand's. This view is still shared

---

[1] *Ber. d. chem. Gesell.* **27**, 2596 (1894).

[2] Blomstrand, *J. prakt. Chem.* **53**, 169 ; **54**, 305 (1896); **55**, 481 (1897), has given a somewhat partial though valuable objective summary of the contending views ; he declares in favor of Bamberger's theory.

by many eminent thinkers, although Hantzsch has rendered the actual structural identity of the salts in question extremely probable by a physico-chemical examination of their behavior.[1]

A complication in the situation next arose by a partial change of front by Bamberger,[2] in which he became convinced that ordinary (normal) diazo compounds contain a *pentavalent* nitrogen atom; and, after advancing a view of his own, which had little in its favor, gravitated toward Blomstrand's formula as the best expression of their properties. He calls the radical $C_6H_5\overset{v}{N}\equiv N$ *phenylazonium*, the radical $C_6H_5\overset{III}{N}=N\text{-}\underset{I}{phenylazo}$; and modifies his views of the isodiazo compounds in so far as to drop the nitrosamine hypothesis and ascribe to these substances the older Kekulé structure.

Hantzsch[3] meanwhile endeavored to strengthen his position by hunting up new cases of isomerism: he succeeded in finding isomeric diazo *cyanides*, which are real and not isocyanides, and which therefore possess the same structure:

$$ClC_6H_4N_2CN$$

Hantzsch, of course, regards these compounds as stereo-isomeric; Bamberger,[4] on the other hand, interprets them on the Blomstrand-Kekulé basis.

Hantzsch's theoretical views now undergo a curious transformation. After declaring that the Blom-

[1] *Ber. d. chem. Gesell.* **27**, 3527 (1894).
[2] *l.c.* **28**, 242, 444 (1895).
[3] *l.c.* **28**, 666 (1895).
[4] *l.c.* **28**, 826 (1895).

strand formula is absolutely incompatible[1] with the behavior of diazo compounds, and that the substances have different structures in the solid state and in solution, he swings around completely and admits[2] with Bamberger that diazo salts with *oxygen* acids are built on the Blomstrand type (for which the name *diazonium* is proposed — this has since been universally adopted).   The same is true of the other salts *in solution:* solutions of diazobenzene chloride, for example, are solutions of phenyldiazonium chloride.   But solid diazobenzene chloride is a syn-diazo compound, and the diazo cyanides are true stereo-isomeres:

$$
\begin{array}{ccc}
C_6H_5N & C_6H_5N & C_6H_5N \\
\| & \| & \| \\
ClN & CN \cdot N & N \cdot CN
\end{array}
$$

Hantzsch thus describes *three* classes of diazo compounds.

The new turn affairs had taken was considerably strengthened by an observation of Bamberger's. When hydroxylamine acts upon nitrosobenzene:

$$C_6H_5NO + H_2NOH = C_6H_5N{=}NOH + H_2O$$

the resulting compound is *isodiazobenzene*.[3]   It may, therefore, be regarded as reasonably certain that this substance has the structure formerly assigned by Kekulé to normal diazobene; and Blomstrand's formula for the normal compound is undisputed victor in the field.

With the exception of a group of substances to be

[1] *Ber. d. chem. Gesell.* **28**, 676 (1895).
[2] *l.c.* 1734.   [3] *l.c.* 1218.

considered in a moment, this is where matters stand to-day. Three years have brought no material change in the situation. Bamberger has given no sign that he adopts his opponent's theories; Hantzsch pursues his goal with steadfast persistence. In part, he has shown that the diazo cyanides do exist in three groups;[1] in addition to the two supposed stereo-isomeres of the phenyl-azo type, he has observed signs of a diazonium cyanide in solution. If a conclusion may be ventured, the honors appear to lie with Hantzsch; the bulk of the evidence seems to be in favor of the stereochemistry of diazobenzene,[2] and Bamberger has maintained a strange silence for nearly two years. But as above remarked, the final word has not yet been said on this perplexing subject.

But while the world is now agreed upon Blomstrand's formula for normal diazonium salts, no agreement has been reached concerning the diazometallic salts. The experimental material upon this phase of the question is extremely complicated, and ill suited to a brief statement. It must suffice to give the contending views, and leave the evidence to further sifting. Bamberger[3] looks upon the isomeric metallic salts of diazobenzene as simple substitution products of normal and isodiazobenzene (according to his formulation):

$$C_6H_5N\diagdown\mkern-8mu\diagup{\!\!\!\!\!N}{\atop ONa} \qquad C_6H_5N=NONa$$

---

[1] *Ber. d. chem. Gesell.* **30**, 2529 (1897).
[2] Cf. Goldschmidt, *l.c.* **28**, 2023 (1895).
[3] *l.c.* **29**, 446 (1896).

Hantzsch,[1] on the other hand, sees in such a compound as the first a chemical impossibility. To him, a real ammonium hydroxide simply *cannot* form such salts; and it must be admitted, the weight of analogy is with him in this contention. His theory of these compounds is that there are no diazonium metal salts, and that the isomerism is to be expressed only by stereochemical formulation:

$$\begin{array}{cc} C_6H_5N & C_6H_5N \\ \parallel & \parallel \\ NaO \cdot N & N \cdot ONa \end{array}$$

But the world has not yet passed the problem in review; and for the present we may believe as we will, or suspend judgment in patient confidence that ere long the truth must prevail.

[1] *l.c.* **28**, 676, 1734 (1895).

AUTHOR INDEX

www.ingramcontent.com/pod-product-compliance
Lightning Source LLC
Chambersburg PA
CBHW030815020726
47499CB00006B/1932